770

AF598452

COMPUTATION
over
FUZZY QUANTITIES

COMPUTATION over FUZZY QUANTITIES

Milan Mareš

Department of Information and Automation Theory
Academy of the Czech Republic
Prague

CRC Press
Boca Raton Ann Arbor London Tokyo

Library of Congress Cataloging-in-Publication Data

Mareš, Milan

Computation over fuzzy quantities / Milan Mareš.

p. cm.

Includes bibliographical references and index.

ISBN 0-8493-7635-1

1. Fuzzy set theory. 2. Computation — fuzzy quantities. I. Mareš, Milan. II. Title.

[DNLM: 1. Fuzzy set theory. QW 710 G289h]

QR749.H64G78 1994

616′.0149—dc20

DNLM/DLC for Library of Congress 94-12345

CIP

Direct all inquiries to CRC Press, Inc., 2000 Corporate Blvd., N.W., Boca Raton, Florida 33431.

International Standard Book Number 0-8493-7635-1

Library of Congress Card Number 94-12345

Printed in the United States of America 1 2 3 4 5 6 7 8 9 0

Printed on acid-free paper

Preface

One of the kindnesses an author can offer to his potential readers is a (preventive) warning concerning the general orientation and formal style of the offered book. The Preface is usually the best place where this can be effectively done.

Since the publication of Zadeh's seminal paper [110] in 1965 the explosion of fuzzy set theoretical papers and monographs indicates a latent call for some alternative concept of the mathematics in general (see [104] or [89], for example) and especially of an uncertainty concept that could be parallel to the probabilistic and generally stochastic approach. The roots of this trend can be seen in qualitatively new connections between reality and its mathematical models, following the large-scale application of highly effective computers. Many new complex phenomena can be processed by computers; their adequate mathematical models are needed. Numerous real phenomena are connected with some kind of uncertainty and not all of them are fully adequate for the classical stochastic description. This situation has already been discussed by numerous authors, and we refer to it only briefly here.

At any rate, the presentation of the fuzzy set theory inspired many authors to use it in a wide scale of applications. It was started by purely practical algorithms, using in some of their steps a numerical evaluation of some possibilities on a scale between 0 and 1. Here fuzzy set theory often only encouraged authors to more sophisticated manipulation with some subjective evaluations. Another part of fuzzy set applications was a purely formal "fuzzification" of abstract mathematical objects, such as category or topology with, at least at its beginning, a very small effect in presenting qualitatively new results.

Somewhere in the middle of the scale of fuzzy set theory applications is the theory of fuzzy numbers and fuzzy quantities. Since these are focused on numerical concepts, they touch on the historical grounds of mathematical thinking. On the other hand, numerical data processing is inherent for many applications, and handling the vagueness hidden in numerical data means handling the vagueness of various practical problems.

It can be easily seen, and it was shown in many works referred to, e.g., in [23] or [49], that vague numbers or vague data cannot be processed by the methods used for exact values. These constraints arise from the natural features of the vagueness and they cannot be fully avoided. It is why much work was done to solve three groups of problems — to suggest effective methods of arithmetic processing of fuzzy quantities, to find the unavoidable limits of their analogy with the arithmetics of exact quantities, and to look for some special conditions under which those limits can be shifted (usually resulting in the weakening of some other desirable properties of the arithmetics).

This book offers a brief view on some results of these efforts. The stress is on the basic methodology and main general ideas — let us say philosophy — of handling fuzzy quantities. The formal results are mostly presented without the proofs which can be found in the referenced papers or easily completed analogously to the proofs of some already known statements. The book does not pretend to be fully exhaustive regarding the wide and still developing area of fuzzy numbers, fuzzy intervals, and fuzzy quantities — the author's ambition was to give a short survey of elementary approaches and their backgrounds. This should give the reader a solid foundation for work in more special branches of the relevant topics.

The text is divided into four main parts. The first contains three chapters devoted to the introduction of the basic concepts. The first chapter reviews the motivation, terminology, and notations of fuzzy set theory. The second introduces the general concept of fuzzy quantities, and the third deals with the fundamental arithmetics of such quantities.

The second part concerns the attempt to find some weakening of the arithmetic operations and relations over fuzzy quantities which could lead to a wider validity of some classical algebraic properties for vague numbers. It is shown that ignoring the existence of some kind of "fuzzy zero" or "fuzzy unit" on one side leads to a weaker form of equality relation but, on the other side, it guarantees the validity of all group properties for fuzzy quantities — of course related to that weaker equality called here the additive, or multiplicative, equivalence.

The general concept of fuzzy quantities investigated in the previous parts can be generalized or even modified without losing its basic properties. Such modifications (to the multidimensional case or to an alternative definition of the arithmetic operations), the specific case of fuzzy numbers and fuzzy intervals with part-wise linear membership functions, and a generalized concept of the fuzziness called $\mathcal{L}$-fuzziness, are briefly mentioned in the third part.

The last and most extensive part is devoted to some potential applications of the presented theory. The particular topics are chosen to illustrate some specific features of different types of applications and the related formal problems. In some cases the main problem of an effective application is to reformulate the original crisp notions in an adequate and mathematically manageable form (as is shown, e.g., in Chapter 13, concerning a fuzzy coalition game), which can be further processed using the previous general results (as, for example, the fuzzy contaminated data in Chapter 10). In other cases the general reformulation allows the studying of some specific modifications partly extending the original model (as it appears in Chapter 12, concerning multicriteria fuzzy decision-making, and its modification in Section 12.6, dealing with the reconstruction of global fuzzy preferences). Also, it is simply interesting to show the methodological consequences of applying the fuzzy approach to some classical arithmetic computations and their interpretations (the fuzzy critical path method is presented in Chapter 11).

The final chapter can be hardly incorporated in the main parts. It contains brief considerations on the structure and different types of fuzziness simultaneously present in fuzzy quantities in general and their characterization by membership functions.

The book also contains a list of relevant literature, mostly cited in the text, an index of the basic concepts, and a list of main notations.

The works on fuzzy set theory and related topics can be divided into two streams. Some of them keep Zadeh's original concept of membership function mapping the basic universum into the closed $[0,1]$ interval. Others suggest some generalizations of this model, mostly substituting for the $[0,1]$ interval a most general lattice

$$\mathcal{L} = (L, \vee, \wedge, \mathbf{0}, \mathbf{1})$$

with the operations $\vee$ (maximum, union, ...) and $\wedge$ (minimum, intersection, ...), and with "minimal" and "maximal" elements (for example [86], [36] and others). The lattice approach is mathematically elegant and general enough to satisfy a great variety of possible images of fuzziness. Nevertheless, for some practical applications of fuzzy set theory, it might seem to be too abstract or distant from immediate needs. We should also accept the fact that in the majority of practical applications the membership function values in $[0,1]$ are the most manageable and preferably used ones.

As this work is assumed to serve as a first orientation, especially for those who are looking for an adequate model of practical fuzzy data in real applications, the following text (with the exception of very brief information on $\mathcal{L}$-fuzziness in Chapter 9) consequently preserves the original simple $[0,1]$ concept of the membership functions. It is not difficult to re-formulate the notions and conclusions presented below into the lattice version, and those who find this approach more acceptable can do so with little effort.

Contents

COMPUTATION over FUZZY QUANTITIES

Part I

Fundamentals

1

Background

This chapter can be omitted by anyone who is familiar with the fundamental concepts of fuzzy set theory. We review here briefly the essential motivation for fuzziness, its main general notions, and some of the notations used in the following parts.

In all following chapters, for real numbers x, y, $x < y$, the symbols $[x, y]$ and (x, y) denote the closed and open interval, respectively. By $\{x, y, z\}$ we denote the set of elements x, y, z.

1.1 Heuristics of Uncertainty

Classical mathematics was fully oriented to the deterministic phenomena of the quantitative component of the world. Even the early probability calculus (that, by the way, appeared surprisingly late) was rather an intellectual toy than an intentional model of reality. Later, the demands of statistical investigations in economy, technical measurements in industry, and the processing of experimental observations in the natural sciences inspired the remarkable development of probabilistic methods. The probability theory did not prove to be fully adequate, however, for all types of uncertainty. It can describe the probability of how many times among 1000 attempts an arrow will strike a target, but it seems to be much more at a loss if it is asked to determine whether the frequency of such hits is "satisfactory" or "very good" or "disreputable". Using the terminology of [49], probability theory is excellent if the *ambiguity* is to be modeled, but its attempt to describe *vagueness* can sometimes be perceived as inconsistent with common sense.

Zadeh's paper [110] and other works by his numerous followers offer a method suitable for the mathematical modeling of vague concepts such as "small", "approximately", "similar", etc., by means of *fuzziness.* In spite of a certain disharmony between "fundamentalists" of both sides

— "probabilists" and "fuzzists" — it is necessary to stress the evident affinity of both approaches. It is quite natural to acknowledge the common uncertainty background of both ambiguity and vagueness. In fact, there is a wide scale of phenomena which can be mathematically modeled by probability as well as by fuzzy set theory. In such cases the structure and orientation of questions to be answered determines which of these two methods should be applied.

Vague numbers and vague quantitative data are examples of phenomena that can be described in both languages, and the choice between them depends on the specificity of each actual problem. This is why this book is intended rather as an alternative approach to the uncertainty hidden in quantitative objects. It aims to demonstrate properties, extent, and limits of the applicability of fuzzy set theory, and also presents methodological questions motivated by the results presented.

1.2 Fuzzy Set

Let us consider a universum $\mathcal{U}$ consisting of some elements. In classical set theory, each subset A of $\mathcal{U}$, symbolically represented $A \subset \mathcal{U}$, can be identified with a characteristic function $\chi_A : \mathcal{U} \to \{0,1\}$ such that for any element $x \in \mathcal{U}$

$$\begin{aligned} \chi_A(x) &= 0 \quad \text{for } x \notin A, \\ &= 1 \quad \text{for } x \in A. \end{aligned}$$

Zadeh [110] generalized this deterministic concept, and suggested expressing the uncertainty of whether an element $x \in \mathcal{U}$ belongs also to A by the values of the characteristic function somewhere over the closed interval $[0,1]$. Then the values of such a modified characteristic function could express the *possibility* that x belongs to A, i.e., that $x \in A$ holds.

To distinguish between fuzzy and *deterministic* concepts (sets, relations, statements, numbers, etc.) it is usual to label the classical deterministic objects *crisp*.

More formally, a *fuzzy subset* A of $\mathcal{U}$, in symbols $A \subset_f \mathcal{U}$, can be identified with a *membership function* $\mu_A : \mathcal{U} \to [0,1]$ such that for any $x \in \mathcal{U}$

$\mu_A(x) = 0$ if x certainly does not belong to A,

$\mu_A(x) = 1$ if x certainly belongs to A,

$\mu_A(x) \in (0,1)$ when there is doubt whether x belongs to A, where the nearer $\mu_A(x)$ is to 1, the higher is the expectation that x belongs to A.

This definition can be generalized if the closed interval $[0,1]$ is replaced with a more general lattice with maximal and minimal elements. In this work we respect the classical $[0,1]$-definition, remembering that the results presented below can, with a certain amount of industry, be transformed into the lattice formalism (a few brief remarks on this topic are presented in Chapter 9).

It is natural to expect that fuzzy sets can be processed in the usual set theory way, namely that set theory operations can be defined over them. Among different approaches to this task, Zadeh's [110] has remained dominant, and it will be used also in this book.

If $A \subset_f \mathcal{U}$, $B \subset_f \mathcal{U}$, then we say that A is a *subset* of B, and write $A \subset B$ iff $\mu_A(x) \leq \mu_B(x)$ for all $x \in \mathcal{U}$.

The *complement* of a fuzzy set $A \subset_f \mathcal{U}$ is also a fuzzy set $\bar{A} \subset_f \mathcal{U}$ such that

$$\mu_{\bar{A}}(x) = 1 - \mu_A(x) \quad \text{for all } x \in \mathcal{U} \tag{1.1}$$

The *union* of two fuzzy sets $A \subset_f \mathcal{U}$, $B \subset_f \mathcal{U}$ is also a fuzzy set $A \cup B \subset_f \mathcal{U}$ with

$$\mu_{A\cup B}(x) = \max\left(\mu_A(x),\, \mu_B(x)\right) \quad \text{for all } x \in \mathcal{U} \tag{1.2}$$

and their *intersection* is a fuzzy set $A \cap B \subset_f \mathcal{U}$ such that

$$\mu_{A\cap B}(x) = \min\left(\mu_A(x),\, \mu_B(x)\right) \tag{1.3}$$

Evidently, the *universum* $\mathcal{U}$ is also a fuzzy subset of itself, namely $\mathcal{U} \subset_f \mathcal{U}$, with

$$\mu_{\mathcal{U}}(x) = 1 \quad \text{for all } x \in \mathcal{U}$$

and, on the other hand, the *empty set* $\emptyset$ is a fuzzy set $\emptyset \subset_f \mathcal{U}$ with

$$\mu_{\emptyset}(x) = 0 \quad \text{for all } x \in \mathcal{U}$$

The concept of a fuzzy set and the operations mentioned above, however common they are, lead to some surprising consequences, discussed in the literature. Let us remember here that the intersection $A \cap \bar{A}$ is not generally empty, and moreover, it is easy to construct $A \subset_f \mathcal{U}$ such that $A \subset \bar{A}$ (simply by putting $\mu_A(x) \leq 0.5$ for all $x \in \mathcal{U}$). When processing fuzzy sets it is very useful to keep in mind these (and other) non-standard properties.

1.3 Fuzzy Relations, Functions, and Operations

Since the concept of a set is present in the roots of many other mathematical notions, it is relatively easy to define their fuzzy modifications.

In the deterministic case a relation ρ over $\mathcal{U}$ can be identified with a subset $S_\rho \subset \mathcal{U} \times \mathcal{U}$ of ordered pairs (x, y) such that the relation $x \rho y$ is fulfilled. Its natural fuzzification means to define the set S_ρ as a fuzzy subset, $S_\rho \subset_f \mathcal{U} \times \mathcal{U}$ with a membership function $\mu_\rho : \mathcal{U} \times \mathcal{U} \to [0,1]$ which represents the *fuzzy relation* ρ over $\mathcal{U}$. Evidently, the value $\mu_\rho(x, y)$ is the degree of possibility that the relation $x \rho y$ is fulfilled.

Analogously, a function φ defined over $\mathcal{U}$ with values in some set $\mathcal{V}$ (usually $\mathcal{V} = R$), i.e., $\varphi : \mathcal{U} \to \mathcal{V}$, can be described in the deterministic case by a set $S_\varphi \subset \mathcal{U} \times \mathcal{V}$ such that for $(x, y) \in S_\varphi$ the equality $y = \varphi(x)$ holds. Then the *fuzzy function* φ can be described by a fuzzy subset $S_\varphi \subset_f \mathcal{U} \times \mathcal{V}$ with a membership function $\mu_\varphi : \mathcal{U} \times \mathcal{V} \to [0,1]$, where $\mu_\varphi(x, y)$ is the possibility that $y = \varphi(x)$. An equivalent approach to the concept of a fuzzy function is to describe for any $x \in \mathcal{U}$ a fuzzy set $S_{\varphi(x)} \subset_f \mathcal{V}$ with $\mu_{\varphi(x)} : \mathcal{V} \to [0,1]$ as a set of possible values of function φ for the input value x.

Finally, an operation ω (such as addition, multiplication, etc.) over $\mathcal{U}$ can also be described deterministically by a set of ordered triples (or generally n-tuples) $S_\omega \subset \mathcal{U} \times \mathcal{U} \times \mathcal{U}$ such that $z = x \,\omega\, y$ for $(x, y, z) \in S_\omega$. In the fuzzy approach, a *fuzzy operation* can be described by a fuzzy subset $S_\omega \subset_f \mathcal{U} \times \mathcal{U} \times \mathcal{U}$ with a membership function $\mu_\omega : \mathcal{U} \times \mathcal{U} \times \mathcal{U} \to [0,1]$, where $\mu_\omega(x, y, z)$ is the possibility that $z = x \,\omega\, y$.

We can simplify some of the considerations given below by noting that there is an essential difference between fuzzy relations, functions, or operations as fuzzy manipulations with crisp objects (elements of $\mathcal{U}$) on one hand, and (in its proper sense) deterministic relations, functions, or operations over fuzzy objects (fuzzy subsets of $\mathcal{U}$), representing their vague inputs on the other. The first case can be represented, for example, by some defined fuzzy operation, say fuzzy addition, of crisp real numbers. The second is essentially a strictly deterministic addition of fuzzy numbers defined in Section 3.1, and used throughout this book.

1.4 Fuzzy Statements

One example of classical objects which can be vague is the *validity* of some statements. Some of them can be not only "*false*" or "*true*", but also "*possible*", "*almost sure*", "*hardly fulfilled*", etc., and all of these expressions represent the fuzzy validity of the relevant statements. If s is a statement then it can be identified with a set of objects for which s is true. Hence, any vague statement s is identified with a fuzzy subset of some universum $\mathcal{U}$ with membership function $\mu_s : \mathcal{U} \to [0,1]$, such that for any $u \in \mathcal{U}$

$$\begin{array}{ll} \mu_s(u) = 1 & \text{means } s \text{ is } \textit{true} \text{ for } u, \\ \mu_s(u) = 0 & \text{means } s \text{ is } \textit{false} \text{ for } u, \\ 0 < \mu_s(u) < 1 & \text{means that } s \text{ is possible for } u \\ & \text{with the degree of possibility } \mu_s(u). \end{array}$$

Fuzzy statements can be processed like any other statements. Using logical operations we can construct new statements whose validity (or possibility) can be derived from the validities of the input statements. Namely, if s, t are fuzzy statements with membership functions (possibilities) μ_s, μ_t, then the *negation* of s is $\neg s$, with

$$\mu_{\neg s}(u) = 1 - \mu_s(u), \quad u \in \mathcal{U} \tag{1.4}$$

the *conjunction* of s and t, denoted $s \wedge t$, fulfills

$$\mu_{s\wedge t}(u) = \min\left(\mu_s(u), \mu_t(u)\right) \tag{1.5}$$

and their *disjunction*, $s \vee t$, fulfills

$$\mu_{s\vee t}(u) = \max\left(\mu_s(u), \mu_t(u)\right) \tag{1.6}$$

The set theory analogies are evident.

An analogous approach can be used even for quantifiers. The possibility that there exists an element u in some set $\mathcal{V} \subset \mathcal{U}$ such that s is fulfilled for u is given by

$$\max\left(\mu_s(u) : u \in \mathcal{V}\right) \quad \text{or} \quad \sup\left(\mu_s(u) : u \in \mathcal{V}\right) \tag{1.7}$$

and the possibility that s is fulfilled for all elements in $\mathcal{V}$ is

$$\min\left(\mu_s(u) : u \in \mathcal{V}\right) \quad \text{or} \quad \inf\left(\mu_s(u) : u \in \mathcal{V}\right) \tag{1.8}$$

If $\mathcal{V}$ is not crisp, but a fuzzy subset of $\mathcal{U}$, i.e., $\mathcal{V} \subset_f \mathcal{U}$ with a membership function $\mu_{\mathcal{V}} : \mathcal{U} \to [0, 1]$, then evidently (1.7) can be replaced with

$$\sup\left(\min\left(\mu_s(u), \mu_{\mathcal{V}}(u)\right) : u \in \mathcal{U}\right) \tag{1.9}$$

(the possibility that there exists $u \in \mathcal{U}$ such that $u \in \mathcal{V}$ and u fulfills s), and (1.8) can be replaced analogously with

$$\inf\left(\min\left(\mu_s(u), \mu_{\mathcal{V}}(u)\right) : u \in \mathcal{U}\right) \tag{1.10}$$

2

Fuzzy Values

One of the potentially vague elements of mathematical models is represented by uncertain quantitative data. They can exist in the form of vague numbers or vague quantities. This book is largely concerned with those fuzzy quantities, and we first formulate their general model.

Throughout this book, we denote by R the set of all real numbers, and by R_0 the set of all non-zero real numbers

$$R_0 = R - \{0\}.$$

2.1 Fuzzy Quantities

By a *fuzzy quantity* we, roughly speaking, mean any vague numerical quantity whose possible values are real numbers. Analogous results can be derived if the set of possible values is a specific subset of R, e.g., the set of natural or rational numbers, which has the necessary algebraic properties. Most of the methods and results presented below remain valid even if we consider fuzzy quantities with possible values in some general group. Of course, the additional properties of such a general group (like its non-commutativity) strongly influence the existence of similar properties of related fuzzy quantities (see [65]).

Here we call a *fuzzy quantity* any fuzzy subset a of R with membership function $\mu_a : R \to [0,1]$ such that

$$\sup\left(\mu_a(x) : x \in R\right) = 1, \tag{2.1}$$

$$\exists x_1^{(a)}, x_2^{(a)} \in R,\ x_1^{(a)} < x_2^{(a)},\ \forall x \in R :$$
$$x \notin \left(x_1^{(a)}, x_2^{(a)}\right) \Rightarrow \mu_a(x) = 0 \tag{2.2}$$

If the support set of μ_a

$$\{x \in R : \mu_a(x) > 0\} \tag{2.3}$$

is finite, then (2.1) becomes

$$\exists x_0 \in R : \mu_a(x_0) = 1$$

and if the support set (2.3) is discrete, then (2.2) implies that it is finite. The set of all fuzzy quantities fulfilling (2.1) and (2.2) will be denoted by

$$\mathbb{R}$$

In fact, none of those conditions is quite necessary for the significant majority of statements presented below. They are natural, well interpretable, they essentially simplify the formalism, notations, and proofs of many important statements derived for fuzzy quantities, and last but not least, they are necessary for the validity of a few of them.

The topics discussed in this book demand the introduction of a few auxiliary notions and notations which simplify the explanation in the following chapters.

If $x \in R$ then we denote by $\langle x \rangle \in \mathbb{R}$ the *degenerated fuzzy quantity*, with

$$\mu_{\langle x \rangle}(x) = 1, \quad \mu_{\langle x \rangle}(y) = 0 \quad \text{for } y \neq x \tag{2.4}$$

If $a \in \mathbb{R}$ then we denote by $-a \in \mathbb{R}$ the *opposite fuzzy quantity*, for which

$$\mu_{-a}(x) = \mu_a(-x) \quad \text{for all } x \in R \tag{2.5}$$

The introduction of *reciprocal quantities* is possible only for fuzzy quantities of non-zero values. Condition (2.2) implies that such non-zero quantities belong to the set $\mathbb{R}_0$, defined by

$$\mathbb{R}_0 = \{a \in \mathbb{R} : \exists \varepsilon > 0,\ \mu_a(x) = 0 \text{ for } x \in (-\varepsilon, \varepsilon)\} \tag{2.6}$$

Thus, for $a \in \mathbb{R}_0$ we define its reciprocal quantity $1/a \in \mathbb{R}$ by

$$\begin{aligned} \mu_{1/a}(x) &= \mu_a(1/x) && \text{for } x \neq 0 \\ &= 0 && \text{for } x = 0 \end{aligned} \tag{2.7}$$

Condition (2.2) and expression (2.6) imply that for $a \in \mathbb{R}_0$ also $1/a \in \mathbb{R}_0$. In those situations in which (2.2) can be omitted, (2.6) can also be reduced to

$$\mathbb{R}_0 = \{a \in \mathbb{R},\ \mu_a(0) = 0\}$$

2.2 Ordering and Extremes

Since fuzzy quantities represent a mathematical model of uncertain numeric quantities, it is desirable to process them analogously to the processing of deterministic (crisp) numbers. Some of the elementary relations over crisp numbers are the relations of equality and strict or weak inequalities. Their fuzzy analogies can be described in the way suggested in this section, but we will see that such analogies are not very strong. Namely, however natural the definitions of "inequality" or "similarity" between fuzzy quantities are, they do not possess some of fundamental properties of ordering. In this sense it is not quite correct to use the term "ordering" here. But even those pseudo-orderings can be used to define the ability of particular fuzzy quantities to be somehow "greater" or "comparable" with others, and to be extreme points of some sets.

First, we should accept a convention. If a, $b \in \mathbb{R}$ then the *equality* symbol $a = b$ always means the point-wise equality of the membership functions, i.e.,

$$\mu_a(x) = \mu_b(x) \quad \text{for all } x \in R$$

Obviously, this strict identity is too strong to represent natural relation between vague quantities, and it is used as an abbreviation of the formal equality of the membership functions. The weaker, and for the nature of the fuzziness more adequate, "*similarity*" relations used below will usually be denoted by the symbol $\sim$ (with some additional index-mark if it is useful). At any rate, we will see in the following sections that the equality symbol $=$ is extremely useful in formulating the membership functions, defining operations over them, and describing their computation methods.

It is necessary to note that the ordering relations between fuzzy quantities are inevitably fuzzy, too. This follows immediately from the fact that general fuzzy quantities gain more values with different degrees of possibility, and those possibilities, together with the ordering relations between particular values, influence the possibility value for the respective ordering between fuzzy quantities as such.

Relations suggested in this section represent only one of the possible approaches to that general problem. The approach is connected with some formal as well as interpretational difficulties, but it is illustrative. The fuzzy ordering and fuzzy ranking methods are among the most intensively developing ones in contemporary fuzzy set theory. Numerous procedures were (and still are) suggested. Some of them can be found in the referenced literature, but their variety shows that none of them is the uniformly best one.

The fuzzy relation "*greater than*", denoted by $\succ$ and defined over $\mathbb{R}$, is the fuzzy subset of $\mathbb{R} \times \mathbb{R}$ with membership function $\mu_\succ : \mathbb{R} \times \mathbb{R} \to [0, 1]$

such that for $a,\, b \in \mathbb{R}$

$$\mu_{\succ}(a,b) = \sup\left(\min(\mu_a(x),\, \mu_b(x)) : x,\, y \in R,\ x > y\right) \tag{2.8}$$

The interpretation of (2.8) is evident if we remember the principles mentioned in section 1.4. The possibility of $a \succ b$ is the possibility that there exists at least one pair of values $x,\, y \in R$ such that $x > y$, and the values of a and b are x and y, respectively.

Analogously, the fuzzy relation "*not smaller than*", denoted by $\succsim$ and defined over $\mathbb{R}$, is the fuzzy subset of $\mathbb{R} \times \mathbb{R}$ with membership function $\mu_{\succsim} : \mathbb{R} \times \mathbb{R} \to [0,1]$ defined for $a,\, b \in \mathbb{R}$ by

$$\mu_{\succsim}(a,b) = \sup\left(\min(\mu_a(x),\, \mu_b(y)) : x,\, y \in R,\ x \geq y\right) \tag{2.9}$$

Formulas (2.8) and (2.9) obviously imply $\mu_{\succsim}(a,b) \geq \mu_{\succ}(a,b)$ for any $a,\, b \in \mathbb{R}$, and in this sense the fuzzy relation $\succ$ is a fuzzy subset of the fuzzy relation $\succsim$.

Finally, the fuzzy relation "*comparable*", denoted by $\sim$ and defined over $\mathbb{R}$, is the fuzzy subset of $\mathbb{R} \times \mathbb{R}$ with membership function $\mu_{\sim} : \mathbb{R} \times \mathbb{R} \to [0,1]$ such that for $a,\, b \in \mathbb{R}$

$$\mu_{\sim}(a,b) = \sup\left(\min(\mu_a(x),\, \mu_b(x)) : x \in R\right) \tag{2.10}$$

Even in this case, it is evident that the comparability relation is a fuzzy subset of the relation "not smaller than", as follows from (2.9) and (2.10).

Moreover, let us consider the disjunction of the statements "being comparable" and "being greater"; this means the union of fuzzy sets (2.9) and (2.10). It is given, for $a,\, b \in \mathbb{R}$, by

$$\begin{aligned}
&\max\left(\mu_{\succ}(a,b),\, \mu_{\sim}(a,b)\right) \\
&= \max\left[\sup\left\{\min(\mu_a(x),\, \mu_b(y)) : x,\, y \in R,\ x > y\right\},\right. \\
&\qquad \left.\sup\left\{\min(\mu_a(x),\, \mu_b(x)) : x \in R\right\}\right] \\
&= \sup\left\{\min(\mu_a(x),\, \mu_b(y)) : x,\, y \in R,\ x \geq y\right\} \\
&= \mu_{\succsim}(a,b)
\end{aligned}$$

and, consequently, the relation $\succsim$ is equivalent with "$\succ$ or $\sim$".

It is easy to verify that, for $a,\, b \in \mathbb{R}$, $a \sim b$ iff $b \sim a$ and, evidently, (2.1) implies $\mu_{\sim}(a,a) = 1$ as well as $\mu_{\succsim}(a,a) = 1$. On the other hand, other usual properties of the relations $\succ$, $\succsim$, and $\sim$, namely their transitivity and the complementarity of $a \succsim b$ and $b \succ a$, are not generally fulfilled. All these facts are illustrated by the following example. It also illustrates the brief note given in the introductory paragraph of this section on the absence of some important properties regarding the ordering relations over fuzzy quantities.

Example 1. Let us consider three fuzzy quantities a, b, $c \in \mathbb{R}$, where

$$\begin{aligned}
&\mu_a(1) = 0.9, \quad \mu_a(3) = 1, \qquad \mu_a(5) = 0.8,\\
&\mu_b(2) = 0.6, \quad \mu_b(3) = 0.7, \quad \mu_b(4) = 1,\\
&\mu_c(5) = 1, \qquad \mu_c(6) = 0.9,
\end{aligned}$$

and $\mu_a(x)$, $\mu_b(y)$, $\mu_c(z)$ vanish in other cases. Then, using (2.8), (2.9), and (2.10), obviously

$$\begin{aligned}
\mu_{\sim}(a,b) &= \min\left(\mu_a(3),\, \mu_b(3)\right) = 0.7\\
\mu_{\succ}(a,b) &= \max\left[\min(\mu_a(3),\, \mu_b(2)), \min(\mu_a(5),\, \mu_b(2)),\right.\\
&\qquad \left.\min(\mu_a(5),\, \mu_b(3)),\, \min(\mu_a(5),\, \mu_b(4))\right]\\
&= \max[0.6,\, 0.6,\, 0.7,\, 0.8] = 0.8
\end{aligned}$$

and analogously

$$\begin{aligned}
\mu_{\succsim}(a,b) &= \max[0.6,\, 0.7,\, 0.6,\, 0.7,\, 0.8] = 0.8\\
\mu_{\succ}(b,a) &= \mu_{\succsim}(b,a) = \max[0.6,\, 0.7,\, 0.7,\, 0.9,\, 1.0] = 1.0
\end{aligned}$$

We can see that there is no complementarity between $\mu_{\succ}(a,b)$ and $\mu_{\succsim}(b,a)$, as $\mu_{\succsim}(a,b) = 0.8 \neq 1 - \mu_{\succ}(b,a) = 0.0$, or inside the triple $\mu_{\succ}(a,b)$, $\mu_{\sim}(a,b)$, $\mu_{\succ}(b,a)$, as the possibility of "$a \succ b$ or $a \sim b$ or $b \succ a$" is (due to (1.5), (1.6)) equal to 0.8 and not to 1.

Similarly,

$$\begin{aligned}
&\mu_{\succ}(a,c) = 0, \quad \mu_{\succsim}(a,c) = 0.8, \quad \mu_{\succ}(c,a) = 1.0, \quad \mu_{\succsim}(c,a) = 1.0,\\
&\mu_{\succ}(b,c) = \mu_{\succsim}(b,c) = 0, \quad \mu_{\succ}(c,b) = \mu_{\succsim}(c,b) = 1,\\
&\mu_{\sim}(a,c) = 0.8, \quad \mu_{\sim}(b,c) = 0
\end{aligned}$$

and no transitivity-like relation between

$$\mu_{\succ}(a,c),\ \mu_{\succ}(c,b),\ \mu_{\succ}(a,b) \quad \text{or} \quad \mu_{\succsim}(a,c),\ \mu_{\succsim}(c,b),\ \mu_{\succsim}(a,b)$$

(or between other similar triples) exists. By "transitivity-like relation" we mean the equality between $\mu_{\succ}(a,b)$ and minimum of $\mu_{\succ}(a,c)$, $\mu_{\succ}(c,b)$ (which means the logical conjunction of $a \succ c$ and $c \succ b$), or similar relations for $\sim$ and $\succsim$.

The previous example illustrates the absence of strong transitivity of the relations $\sim$, $\succ$, or $\succsim$, but we can also consider the possibility of its modified weaker form. For example, if a, b, $c \in \mathbb{R}$, then we may ask if the possibility of the relation $a \succ b$ (or $a \sim b$, $a \succsim b$) is not at least as large as the possibility that $a \succ c$ and $c \succ b$; in our notation, if

$$\mu_{\succ}(a,b) \geq \min(\mu_{\succ}(a,c),\, \mu_{\succ}(b,c))$$

(and analogously for $\mu_{\succsim}(a,b)$ and $\mu_{\sim}(a,b)$). The values of possibilities calculated in the previous example do not exclude this relation, but its validity is not general, as shown below.

Example 2. Let a, b, $c \in \mathbb{R}$ be such that

$$\mu_a(2) = 0.3, \quad \mu_a(3) = 0.5, \quad \mu_a(4) = 0.7, \quad \mu_a(5) = 1.0,$$
$$\mu_b(1) = 1.0, \quad \mu_b(2) = 0.7, \quad \mu_b(3) = 0.5, \quad \mu_b(4) = 0.3,$$
$$\mu_c(1) = \mu_c(2) = \mu_c(4) = \mu_c(5) = 0.8, \quad \mu_c(3) = 1.0,$$
$$\mu_a(x) = 0, \quad \mu_b(x) = 0, \quad \mu_c(x) = 0 \quad \text{for other } x \in R.$$

Then, using (2.8), (2.9), (2.10), it is easy to calculate

$$\mu_{\sim}(a,b) = 0.5, \quad \mu_{\sim}(a,c) = 0.8, \quad \mu_{\sim}(c,b) = 0.8,$$
$$\mu_{\succsim}(b,a) = 0.5, \quad \mu_{\succsim}(b,c) = 0.8, \quad \mu_{\succsim}(c,a) = 0.8,$$
$$\mu_{\succ}(b,a) = 0.3, \quad \mu_{\succ}(b,c) = 0.7, \quad \mu_{\succ}(c,a) = 0.7.$$

This means that $\mu_{\sim}(a,b)$ is less than $\min(\mu_{\sim}(a,c),\, \mu_{\sim}(c,b))$, and similarly for the remaining two triples.

The lack of transitivity means that the relations $\succ$ or $\succsim$ are not ordering relations and $\sim$ is not equivalence in the usual sense. They can, however, be used to define certain kinds of extremality over fuzzy quantities.

As the relation "not greater than" is fuzzy, it is evident that also the extremes (maximum and minimum) of some set of fuzzy quantities are represented by its fuzzy subsets. Let $\mathbb{A} \subset \mathbb{R}$ be a (non-empty) set of fuzzy quantities. Its maximum $\mathbb{A}_{\max}$ is a fuzzy subset of $\mathbb{A}$ with membership function $\mu_{\max} : \mathbb{A} \to [0,1]$ such that for any $a \in \mathbb{A}$

$$\mu_{\max}(a) = \inf\left(\mu_{\succsim}(a,b) : b \in \mathbb{A}\right) \tag{2.11}$$

which formula denotes the possibility that a is not smaller than any $b \in \mathbb{A}$. Analogously, the minimum of $\mathbb{A}$, denoted by $\mathbb{A}_{\min}$, is a fuzzy subset of $\mathbb{A}$ with membership function $\mu_{\min} : \mathbb{A} \to [0,1]$ such that

$$\mu_{\min}(a) = \inf\left(\mu_{\succsim}(b,a) : b \in \mathbb{A}\right) \tag{2.12}$$

for any $a \in \mathbb{A}$. This denotes the possibility that no b in $\mathbb{A}$ is smaller than a.

Example 3. Let us consider $\mathbb{A} = \{a, b, c\}$, where the fuzzy quantities are described in Example 1. Then, using the values derived in Example 1,

$$\mu_{\max}(a) = \min\left[\mu_{\succsim}(a,b),\ \mu_{\succsim}(a,c),\ \mu_{\succsim}(a,a)\right] = \min[0.8,\ 0,\ 1] = 0$$
$$\mu_{\max}(b) = \min[1,\ 0,\ 1] = 0$$
$$\mu_{\max}(c) = \min[1,\ 1,\ 1] = 1$$

and

$$\mu_{\min}(a) = \min[1,\ 1,\ 1] = 1$$
$$\mu_{\min}(b) = \min[0.8,\ 1,\ 1] = 0.8$$
$$\mu_{\min}(c) = \min[0,\ 0,\ 1] = 0$$

2.3 Arithmetic Operations

In the rest of this book we focus on arithmetic operations over the set $\mathbb{R}$. There exists a general method of calculating the membership function of the output from the membership functions of the input fuzzy quantities. It is known as the *representation principle* [110], [23]. For our purpose we recall it in a rather specialized form, corresponding with the essential structure of real-valued fuzzy quantities.

Let $f : R \times R \to R$ be a binary operation over real numbers. Then it can be extended to the operation over the set $\mathbb{R}$ of fuzzy quantities. If we denote for $a,\ b \in \mathbb{R}$ the quantity $c = f(a,b)$, then the membership function μ_c is derived from the membership functions μ_a and μ_b by

$$\mu_c(z) = \sup\left[\min(\mu_a(x),\ \mu_b(y)) : x,\ y \in R,\ z = f(x,y)\right] \tag{2.13}$$

for any $z \in R$.

The interpretation of (2.13) is obvious if we use the principles described in section 1.4. The possibility that the fuzzy quantity $c = f(a,b)$ achieves value $z \in R$ is as great as the most possible combination of real numbers x, y such that $z = f(x,y)$, where the values of a and b are x and y, respectively.

It is not difficult to see that formulas (2.8), (2.9), and (2.10) also reflect the representation principle, used in rather generalized form.

In the following chapters we deal mostly with $f(x,y) = x + y$ and $f(x,y) = x \cdot y$, but in a few sections also with exponents.

3

Arithmetics of Fuzzy Quantities

The representation principle (2.13) can be applied to the usual arithmetic operations, such as addition or multiplication. Extended from crisp quantities to fuzzy ones, they preserve only some of their classical properties [23], [64], [67]. In this chapter we review the validity of particular algebraic rules for fuzzy quantities, and briefly discuss the existing differences between the crisp and fuzzy cases.

3.1 Addition

If a, $b \in \mathbb{R}$ are fuzzy quantities with membership functions μ_a, μ_b, respectively, then the fuzzy quantity $a \oplus b$ with

$$\mu_{a\oplus b}(x) = \sup_{y \in R} (\min(\mu_a(y), \mu_b(x-y))) \tag{3.1}$$

is called a *sum of a and b*. Formula (3.1) is obviously an application of the representation principle (2.13), where $f(x, y) = x + y$.

It can easily be seen that $a \oplus b$ fulfills (2.1) and (2.2) if a and b do so. Moreover

$$\mu_{a\oplus b}(x) = \sup_{z \in R} (\min(\mu_a(x-z), \mu_b(z))) \tag{3.2}$$

as can be derived from (3.1) by substituting $y = x - z$, and for any $y \in R$ and $\langle y \rangle \in \mathbb{R}$ (see (2.4)), formula (3.1) implies

$$\mu_{a\oplus\langle y\rangle}(x) = \mu_a(x-y) \tag{3.3}$$

To illustrate formulas (3.1) and (3.2) and their practical effect we introduce a few numerical examples. The simplest types of fuzzy quantities are those with discrete finite support.

Example 4. Let us consider fuzzy quantities $a \in \mathbb{R},\ b \in \mathbb{R}$ such that

$$\mu_a(1) = \tfrac{1}{3}, \quad \mu_a(2) = \tfrac{2}{3}, \quad \mu_a(3) = 1.0, \quad \mu_a(x) = 0 \quad \text{for other } x \in R,$$
$$\mu_b(4) = \tfrac{1}{2}, \quad \mu_b(5) = 1.0, \quad \mu_b(6) = \tfrac{1}{2}, \quad \mu_b(x) = 0 \quad \text{for other } x \in R.$$

Then, using (3.1), we can calculate that

$$\mu_{a\oplus b}(5) = \min(\mu_a(1),\ \mu_b(4)) = \min\left(\frac{1}{3}, \frac{1}{2}\right) = \frac{1}{3}$$

$$\begin{aligned}\mu_{a\oplus b}(6) &= \max\left[\min(\mu_a(1),\ \mu_b(5)), \min(\mu_a(2),\ \mu_b(4))\right] \\ &= \max\left[\min\left(\frac{1}{3}, 1\right), \min\left(\frac{2}{3}, \frac{1}{2}\right)\right] = \max\left(\frac{1}{3}, \frac{1}{2}\right) = \frac{1}{2}\end{aligned}$$

and analogously

$$\mu_{a\oplus b}(7) = \max\left(\frac{1}{3}, \frac{2}{3}, \frac{1}{2}\right) = \frac{2}{3}, \quad \mu_{a\oplus b}(8) = \max\left(\frac{1}{2}, 1\right) = 1$$

$$\mu_{a\oplus b}(9) = \min(\mu_a(3),\ \mu_b(6)) = \min\left(1, \frac{1}{2}\right) = \frac{1}{2}$$

$\mu_{a\oplus b}(x) = 0$ for other $x \in R$.

The sum of discrete and continuous fuzzy quantities, considered for illustration in the following example, can also be calculated without serious difficulties.

Example 5. Let $a,\ b \in \mathbb{R}$ be fuzzy quantities with membership functions

$$\begin{aligned}\mu_a(y) &= y - 1 && \text{for } y \in [1,2], && \mu_b(4) = 1 \\ &= 3 - y && \text{for } y \in [2,3], && \mu_b(5) = \tfrac{1}{2}, \\ &= 0 && \text{for } y \notin (1,3), && \mu_b(z) = 0, \quad \text{for } z \notin \{4,5\}.\end{aligned}$$

Then for any $x < 5$, $\mu_a(x - z) = 0$ and for any $x > 8$ also $\mu_a(x - z) = 0$ for $z \in \{4,5\}$. This means that

$$\mu_{a\oplus b}(x) = 0 \quad \text{for } x \notin [5,8]$$

as follows from (3.2).

If $x \in [5,6]$, then for $z = 5$, $\mu_a(x-z) = 0$ and for $z = 4$ $\mu_a(x-z) = \mu_a(x-4) = x - 4 - 1 = x - 5$. Hence,

$$\mu_{a\oplus b}(x) = \sup_{z\in R}\left(\min(\mu_a(x - z),\ \mu_b(z))\right)$$

$$= \min(\mu_a(x-4),\, \mu_b(4)) = \min(x-5,\, 1) = x-5$$

If $x \in [6,\, 6.5]$ then

$$\mu_a(x-z) = 3 - x + z \quad \text{for } z = 4,$$
$$\mu_a(x-z) = x - z - 1 \quad \text{for } z = 5$$

and consequently, using (3.2),

$$\mu_{a \oplus b}(x) = \max\left[\min(7-x,\, 1),\, \min\left(x-6,\, \frac{1}{2}\right)\right]$$
$$= \max[7-x,\, x-6] = 7-x$$

If $x \in [6.5,\, 7]$ then analogously

$$\mu_a(x-z) = 3 - x + z = 7 - x \quad \text{for } z = 4,$$
$$= x - z - 1 = x - 6 \quad \text{for } z = 5,$$

and again

$$\mu_{a \oplus b}(x) = \max\left[\min(7-x,\, 1),\, \min\left(x-6,\, \frac{1}{2}\right)\right]$$
$$= \max\left[7-x,\, \frac{1}{2}\right] = \frac{1}{2}$$

For $x \subset [7, 8]$ $\mu_a(x-z) = 0$ for $z = 4$,
$\mu_a(x-z) = 3 - (x-z)$ for $z = 5$.

This means that

$$\mu_{a \oplus b}(x) = \max\left[0,\, \min\left(8-x,\, \frac{1}{2}\right)\right]$$
$$= \min\left(8-x,\, \frac{1}{2}\right) \begin{array}{l} = \frac{1}{2} \quad \text{for } x \in [7,\, 7.5] \\ = 8-x \text{ for } x \in [7.5,\, 8] \end{array}$$

Summarizing the previous calculations, we can see that $\mu_{a \oplus b}$ is a continuous function such that

$$\begin{aligned} \mu_{a \oplus b}(x) &= 0 && \text{for } x \le 5 \text{ and } x \ge 8 \\ &= x-5 && \text{for } x \in [5, 6] \\ &= 7-x && \text{for } x \in [6,\, 6.5] \\ &= \tfrac{1}{2} && \text{for } x \in [6.5,\, 7.5] \\ &= 8-x && \text{for } x \in [7.5,\, 8] \end{aligned}$$

If both membership functions μ_a, μ_b are continuous then the computation of $\mu_{a\oplus b}$ is more complex. The following three examples can illustrate some of the basic features of the operation of addition. The first example is relatively simple.

Example 6. Let the membership functions of fuzzy quantities a and b be

$$\begin{aligned} \mu_a(y) &= y-1 && \text{for } y \in [1,2], \\ &= 0 && \text{for } y \notin [1,2], \end{aligned}$$

$$\begin{aligned} \mu_b(z) &= 4-z && \text{for } z \in [3,4], \\ &= 0 && \text{for } z \notin [3,4]. \end{aligned}$$

Using (3.1), we can derive the values of $\mu_{a\oplus b}(x)$, $x \in R$. For $x < 4$, $\mu_{a\oplus b}(x) = 0$ as at least one of the values $\mu_a(y)$ and $\mu_b(x-y)$ in (3.1) is equal to 0.

For $x \in [4,5]$ we can use (3.2) and obtain

$$\sup_{z\in R}(\min(\mu_a(x-z),\, \mu_b(z))) = \sup_{z\in[3,4]}(\min(\mu_a(x-z),\, \mu_b(z))$$
$$= \min(\mu_a(x-3),\, \mu_b(3)) = \mu_a(x-3) = x-4$$

For $x \in [5,6]$ we use (3.1) to find

$$\sup_{y\in R}(\min(\mu_a(y),\, \mu_b(x-y))) = \sup_{y\in[1,2]}(\min(\mu_a(y),\, \mu_b(x-y)))$$
$$= \min(\mu_a(2),\, \mu_b(x-2)) = \mu_b(x-2) = 6-x$$

For $x > 6$ at least one of the values $\mu_a(y)$ and $\mu_b(x-y)$ in (3.1) is equal to 0 and consequently $\mu_{a\oplus b}(x) = 0$. Function $\mu_{a\oplus b}$ is evidently continuous, and $\mu_{a\oplus b}(4) = \mu_{a\oplus b}(6) = 0$, $\mu_{a\oplus b}(5) = 1$.

In the following example we compute the sum of two fuzzy quantities, which are in a certain sense symmetric.

Example 7. Let us consider $a,\, b \in \mathbb{R}$ such that

$$\begin{aligned} \mu_a(y) &= y - 1 \quad \text{for } y \in [1,2], \\ &= 3 - y \quad \text{for } y \in [2,3], \\ &= 0 \qquad\quad \text{for } y \notin [1,3], \end{aligned}$$

$$\begin{aligned} \mu_b(z) &= 5 - z \quad \text{for } z \in [4,5], \\ &= z - 5 \quad \text{for } z \in [5,6], \\ &= 0 \qquad\quad \text{for } z \notin [4,6]. \end{aligned}$$

Then $\mu_{a\oplus b}(x) = 0$ for $x < 5$ as for any $y \in R$ at least one of the values $\mu_a(y),\ \mu_b(x-y)$ in (3.1) vanishes.

For $x \in [5,6]$ and for any $y \in [2,3]$ is $\mu_b(x-y) = 0$. Hence, for $x \in [5,6]$

$$\sup_{y\in R}\left(\min(\mu_a(y),\ \mu_b(x-y))\right) = \sup_{y\in[1,2]}\left(\min(\mu_a(y),\ \mu_b(x-y)\right)$$

$$= \min(\mu_a(x-4),\ \mu_b(4)) = \mu_a(x-4) = x - 5$$

For $x \in [6,8]$ formula (3.1) turns into

$$\sup_{y\in R}(\min(\mu_a(y),\ \mu_b(x-y))$$

$$= \max\left[\sup_{y\in[1,2]}\left(\min(\mu_a(y),\ \mu_b(x-y)\right),\right.$$

$$\left.\sup_{y\in[2,3]}\left(\min(\mu_a(y),\ \mu_b(x-y))\right)\right]$$

where

$$\sup_{y\in[1,2]}\left(\min(\mu_a(y),\ \mu_b(x-y))\right) = \max_{y\in[1,2]}\left(\min(y-1,\ x-y-5)\right)$$

$$\sup_{y\in[2,3]}\left(\min(\mu_a(y),\ \mu_b(x-y))\right) = \max_{y\in[2,3]}\left(\min(3-y,\ 5-x+y)\right)$$

The first of these maxima is for any $x \in [6,8]$ achieved for such $y \in [1,2]$ for which

$$y - 1 = x - y - 5, \quad \text{i.e.,} \quad y = \frac{x}{2} - 2.$$

Substituting this value into $\mu_a(y)$ or $\mu_b(x-y)$ we obtain

$$\max_{y\in[1,2]}\left(\min(y-1,\ x-y-5)\right) = \mu_a(y) = \mu_b(x-y) = \frac{x}{2} - 3.$$

Analogously, the second one of those maxima is for any $x \in [6,8]$ achieved for the $y \in [2,3]$ for which $\mu_a(y) = \mu_b(x-y)$, i.e.,

$$3 - y = 5 - x + y, \quad \text{which means} \quad y = \frac{x}{2} - 1$$

Substituting this value into $\mu_a(y)$ or $\mu_b(x-y)$, we obtain

$$\max_{y\in[2,3]} (\min(3-y,\, 5-x+y)) = 4 - \frac{x}{2}$$

Regarding the decomposition of the supremum in (3.1) into maximum of two suprema we get

$$\mu_{a\oplus b}(x) = \max\left(\frac{x}{2} - 3,\, 4 - \frac{x}{2}\right)$$

It means that for $x \in [6,8]$

$$\begin{aligned} \mu_{a\oplus b}(x) &= 4 - \tfrac{x}{3} && \text{if } x \in [6,7] \\ &= \tfrac{x}{2} - 3 && \text{if } x \in [7,8] \end{aligned}$$

The computation of $\mu_{a\oplus b}(x)$ for $x \in [8,9]$ is similar to the procedure used for $x \in [5,6]$. For any $y \in [1,2]$ the equality $\mu_b(x-y) = 0$ holds and consequently

$$\begin{aligned} \mu_{a\oplus b}(x) &= \sup_{y\in[2,3]} (\min(\mu_a(y),\, \mu_b(x-y))) \\ &= \min(\mu_a(x-6),\, \mu_b(6)) = \mu_a(x-6) = 9 - x \end{aligned}$$

Evidently, for $x > 9$ and any $y \in [1,3]$ the value of $\mu_b(x-y)$ vanishes, which means that $\mu_{a\oplus b}(x) = 0$.

Summarizing the above calculations, we obtain

$$\begin{aligned} \mu_{a\oplus b}(x) &= 0 && \text{for } x \notin (5,9), \\ &= x - 5 && \text{for } x \in [5,6], \\ &= 4 - \tfrac{x}{2} && \text{for } x \in [6,7], \\ &= \tfrac{x}{2} - 3 && \text{for } x \in [7,8], \\ &= 9 - x && \text{for } x \in [8,9]. \end{aligned}$$

The membership function $\mu_{a\oplus b}$ is continuous, in a certain sense symmetric, and $\mu_{a\oplus b}(5) = \mu_{a\oplus b}(9) = 0,\ \mu_{a\oplus b}(6) = \mu_{a\oplus b}(8) = 1,\ \mu_{a\oplus b}(7) = 0.5$.

By the next illustrative example we show the changes of the sum $a \oplus b$ caused by a slight change of one of its components. It is useful to compare this result with the result of Example 7.

Example 8. In this example we consider the fuzzy quantity $b \in \mathbb{R}$, which was used in the previous example, i.e.,

$$\begin{aligned} \mu_b(z) &= 5 - z \quad \text{for } z \in [4,5], \\ &= z - 5 \quad \text{for } z \in [5,6], \\ &= 0 \quad \text{for } z \notin [4,6], \end{aligned}$$

and the fuzzy quantity $a \in \mathbb{R}$ such that

$$\begin{aligned} \mu_a(y) &= y - 1 \quad \text{for } y \in [1,2], \\ &= 1 \quad \text{for } y \in [2,3), \\ &= 0 \quad \text{for } y \notin [1,3). \end{aligned}$$

Then it is easy to derive, similarly to the previous case, that

$$\mu_{a \oplus b}(x) = 0 \quad \text{for} \quad x < 5 \text{ or } x \geq 9$$

as for such $x \notin [5,9)$ and for any $y \in [1,3)$ $\mu_b(x-y) = 0$. If $x \in [5,6]$ then, analogously to Example 7,

$$\mu_{a \oplus b}(x) = x - 5$$

If $x \in [6,7]$ then

$$\sup_{z \in R}(\min(\mu_a(x-z),\, \mu_b(z)))$$

$$= \max\left[\sup_{z \in [5,6]}(\min(\mu_a(x-z),\, \mu_b(z))), \min(\mu_a(x-4),\, \mu_b(4))\right]$$

As $\mu_b(4) = \mu_a(x-4) = 1$ for $x \in [6,7]$ the maximum given above must be equal to 1.

For $x \in [7,8]$ similarly

$$\sup_{y \in R}(\min(\mu_a(y),\, \mu_b(x-y)))$$

$$= \max\left[\max_{y \in [1,2]}(\min(\mu_a(y),\, \mu_b(x-y))),\, \min(\mu_a(3),\, \mu_b(x-3))\right]$$

Here

$$\min(\mu_a(3),\, \mu_b(x-3)) = \min(1,\, 8-x) = 8-x$$

and in the formula

$$\max_{y \in [1,2]}(\min(\mu_a(y),\, \mu_b(x-y)))$$

the desired maximum is achieved for $y \in [1,2]$, for which

$$\mu_a(y) = \mu_b(x-y), \quad \text{i.e.,} \quad y - 1 = x - y - 5$$

The equality is fulfilled for

$$y = \frac{x}{2} - 2$$

for which the maximum value is

$$\mu_a(y) = \mu_b(x-y) = \frac{x}{2} - 3$$

This means that for $x \in [7,8]$

$$\mu_{a\oplus b}(x) = \max\left(8 - x, \frac{x}{2} - 3\right)$$

i.e.,

$$\begin{aligned} \mu_{a\oplus b}(x) &= 8 - x \quad \text{for } x \in \left[7,\, 7\tfrac{1}{3}\right] \\ &= \tfrac{x}{2} - 3 \quad \text{for } x \in \left[7\tfrac{1}{3},\, 8\right] \end{aligned}$$

Finally, for $x \in [8,9)$,

$$\sup_{z \in R}(\min(\mu_a(x-z),\, \mu_b(z))) = \min(\mu_a(x-5),\, \mu_b(5)) = 1$$

Summarizing the previous results we obtain

$$\begin{aligned} \mu_{a\oplus b}(x) &= x - 5 && \text{for } x \in [5,6], \\ &= 1 && \text{for } x \in [6,7], \\ &= 8 - x && \text{for } x \in \left[7,\, 7\tfrac{1}{3}\right], \\ &= \tfrac{x}{2} - 3 && \text{for } x \in \left[7\tfrac{1}{3},\, 8\right], \\ &= 1 && \text{for } x \in [8,9), \\ &= 0 && \text{for } x \leq 5 \text{ or } x \geq 9. \end{aligned}$$

The membership function $\mu_{a\oplus b}(x)$ is discontinuous for $x = 9$, and it is not symmetric. Let us complete the marginal values of intervals: $\mu_{a\oplus b}(6) = 1 = \mu_{a\oplus b}(7)$, $\mu_{a\oplus b}\left(\frac{22}{3}\right) = $
$= \frac{2}{3}$, $\mu_{a\oplus b}(8) = 1$.

The examples above, numerically solved, offer the opportunity to compare the results obtained with the behavior of the same fuzzy quantities in other types of operations, described in the following sections.

The operation $\oplus$ is commutative and associative; this means that for any a, b, $c \in \mathbb{R}$

$$a \oplus b = b \oplus a \tag{3.4}$$

$$a \oplus (b \oplus c) = (a \oplus b) \oplus c \tag{3.5}$$

as shown, for example, in [64]. The commutativity follows immediately from (3.1) and (3.2). The associativity can be proved by simple substitution:

$$\begin{aligned}
\mu_{(a\oplus b)\oplus c}(u) &= \sup_{z\in R}\left(\min\left(\mu_{a\oplus b}(z),\, \mu_c(u-z)\right)\right) \\
&= \sup_{z\in R}\left(\min\left(\sup_{x\in R}\left[\min(\mu_a(x),\, \mu_b(z-x))\right],\, \mu_c(u-z)\right)\right) \\
&= \sup_{z\in R}\left(\sup_{x\in R}\left[\min\left(\min(\mu_a(x),\, \mu_b(z-x)),\, \mu_c(x)\right)\right]\right) \\
&= \sup_{x\in R}\left(\min\left(\mu_a(x),\, \sup_{z\in R}\left[\min(\mu_b(z-x), \mu_c(u-z))\right]\right)\right) \\
&= \sup_{x\in R}\left(\min\left(\mu_a(x),\, \sup_{y\in R}\left[\min(\mu_b(y),\, \mu_c(u-x-y))\right]\right)\right) \\
&= \sup_{x\in R}\left(\min(\mu_a(x),\, \mu_{b\oplus c}(u-x))\right) = \mu_{a\oplus(b\oplus c)}(u)
\end{aligned}$$

where y is substituted for $z-x$. Formula (3.3) implies for any $a \in \mathbb{R}$

$$a \oplus \langle 0\rangle = a \tag{3.6}$$

which means that the set $\mathbb{R}$ with operation $\oplus$ is a *commutative additive monoid.*

Since the addition of crisp real numbers is a group operation, it would be pleasant to find the same group properties for $\oplus$ over $\mathbb{R}$. In accordance with the validity of (3.4), (3.5), and (3.6), only the existence of the opposite element is missing. For any $a \in \mathbb{R}$ the fuzzy quantity $-a$ given by (2.5) seems to be its natural opposite. Unfortunately, there is no general equality between $a \oplus (-a)$ and $\langle 0\rangle$, which is the zero-element with respect to (3.6).

Example 9. Let $a \in \mathbb{R}$ be such that

$$\mu_a(-1) = 1, \quad \mu_a(2) = 0.5, \quad \mu_a(x) = 0 \quad \text{for } x \notin \{-1, 2\}$$

Then $-a \in \mathbb{R}$ is such that

$$\mu_{-a}(-2) = 0.5, \quad \mu_{-a}(1) = 1, \quad \mu_{-a}(x) = 0 \quad \text{for } x \notin \{-2, 1\}$$

and by (3.1)

$$\begin{aligned}
&\mu_{a\oplus(-a)}(-3) = 0.5, \quad \mu_{a\oplus(-a)}(0) = 1, \quad \mu_{a\oplus(-a)}(3) = 0.5, \\
&\mu_{a\oplus(-a)}(x) = 0 \quad \text{for other } x \in R.
\end{aligned}$$

One conclusion following immediately from (3.1), (2.1), and (2.5) is that $\mu_{a\oplus(-a)}(0) = 1$ for any $a \in \mathbb{R}$.

3.2 Multiplication

Analogously to the previous case, the multiplication of fuzzy quantities can also be defined, using the representation principle (2.13).

The definition for fuzzy quantities with non-zero values from $\mathbb{R}_0$ (see Equation 2.6) is a special form of the representation principle (2.13) closely analogous to (3.1). If $a, b \in \mathbb{R}_0$, then the fuzzy quantity $a \odot b$ with

$$\mu_{a\odot b}(x) = \sup_{y\in R_0} (\min(f_a(y), f_b(x/y))) \tag{3.7}$$

for any $x \in R_0$, and

$$\mu_{a\odot b}(0) = 0$$

is called the *product of* a *and* b. This definition can be extended to the whole set $\mathbb{R}$ as suggested in [23], so that for any $a, b \in \mathbb{R}$, $\mu_{a\odot b}(x)$ is given by (3.7) for $x \neq 0$, and

$$\mu_{a\odot b}(0) = \max(\mu_a(0), \mu_b(0)) \tag{3.8}$$

This definition of $\mu_{a\odot b}(0)$ is natural. The value of the product is 0, with the possibility that the value of at least one input factor is equal to 0. It can easily be verified that $a \odot b$ fulfills (2.1) and (2.2) if a and b fulfill (2.1) and (2.2).

Also analogously to (3.2), for $x \neq 0$

$$\mu_{a\odot b}(x) = \sup_{z\neq 0} (\min(\mu_a(x/z), \mu_b(z)))$$

which equality can be derived from (3.7) by substituting x/z for $y \in R_0$.

If $r \in R$ is a real number and $a \in \mathbb{R}$, then it is useful to define the *crisp product of* a *and* r as a fuzzy quantity $r \cdot a \in \mathbb{R}$ such that

$$\begin{aligned} \mu_{r\cdot a}(x) &= \mu_a(x/r) \quad \text{for } r \neq 0 \\ &= \mu_{\langle 0\rangle}(x) \quad \text{for } r = 0 \end{aligned} \tag{3.9}$$

Relations (2.4), (3.7), and (3.9) imply that

$$r \cdot a = \langle r \rangle \odot a. \tag{3.10}$$

To prepare formal tools for dealing with products of fuzzy quantities as well as with other related concepts, it will be useful to introduce the following notations.

$$\mathbb{R}^+ = \{a \in \mathbb{R}_0 : \forall x < 0,\ \mu_a(x) = 0\}$$
$$\mathbb{R}^- = \{a \in \mathbb{R}_0 : \forall x > 0,\ \mu_a(x) = 0\}$$
$$\mathbb{R}^* = \mathbb{R}^+ \cup \mathbb{R}^-$$

The fuzzy quantities from $\mathbb{R}^+$, $\mathbb{R}^-$, and $\mathbb{R}^*$ are called *positive, negative,* and *signed*, respectively. It is easy to verify the following properties.

If $a,\ b \in \mathbb{R}^+,\ r_1,\ r_2 \in R,\ r_1 < 0 < r_2$ then $a \odot b \in \mathbb{R}^+,\ r_1 \cdot a \in \mathbb{R}^-,\ r_2 \cdot a \in \mathbb{R}^+,\ a \oplus b \in \mathbb{R}^+,\ 1/a \in \mathbb{R}^+,\ -a \in \mathbb{R}^-$.

If $a,\ b \in \mathbb{R}^-,\ r_1,\ r_2 \in R,\ r_1 < 0 < r_2$ then $a \odot b \in \mathbb{R}^+,\ r_1 \cdot a \in \mathbb{R}^+,\ r_2 \cdot a \in \mathbb{R}^-,\ a \oplus b \in \mathbb{R}^-,\ (1/a) \in \mathbb{R}^-,\ -a \in \mathbb{R}^+$.

If $a \in \mathbb{R}^+,\ b \in \mathbb{R}^-$ then $a \odot b \in \mathbb{R}^-$.

If $a \in \mathbb{R}_0 - \mathbb{R}^*$ then $(-a)$, $(1/a)$ and $r \cdot a$ also belong to $\mathbb{R}_0 - \mathbb{R}^*$ for any $r \in R_0$.

Since the fuzzy quantities used in Examples 4 – 8 in the previous section belong to $\mathbb{R}_0$ and, moreover, to $\mathbb{R}^+$, it is also possible to calculate their products. The comparison with the results of addition can be interesting, too.

Example 10. Let us consider fuzzy quantities $a,\ b \in \mathbb{R}$ with discrete support sets,

$$\begin{aligned}
&\mu_a(1) = \tfrac{1}{3}, && \mu_b(4) = \tfrac{1}{2},\\
&\mu_a(2) = \tfrac{2}{3}, && \mu_b(5) = 1,\\
&\mu_a(3) = 1, && \mu_b(6) = \tfrac{1}{2},\\
&\mu_a(y) = 0 \quad \text{for } y \notin \{1,2,3\}, && \mu_b(z) = 0 \quad \text{for } z \notin \{4,5,6\}.
\end{aligned}$$

Using (3.7) we derive

$$\mu_{a\odot b}(4) = \min(\mu_a(1),\ \mu_b(4)) = \frac{1}{3}$$

and analogously

$$\begin{aligned}
&\mu_{a\odot b}(5) = \tfrac{1}{3}, \quad \mu_{a\odot b}(6) = \tfrac{1}{3}, \quad \mu_{a\odot b}(8) = \tfrac{1}{2},\\
&\mu_{a\odot b}(10) = \tfrac{2}{3}, \quad \mu_{a\odot b}(12) = \tfrac{1}{2}, \quad \mu_{a\odot b}(15) = 1,\\
&\mu_{a\odot b}(18) = \tfrac{1}{2}, \quad \mu_{a\odot b}(x) = 0 \quad \text{for other } x \in R\ .
\end{aligned}$$

We can note a certain loss of local monotonicity, which existed in the additive case.

The product of continuous and discrete fuzzy quantities is in a certain sense similar to their sum with respect to continuity and local monotonicity, as illustrated in Example 5 and in the next example.

Example 11. The product of fuzzy quantities considered in Example 5, i.e., a, $b \in \mathbb{R}$ such that

$$\begin{aligned} \mu_a(y) &= y-1 \quad \text{for } y \in [1,2], \\ &= 3-y \quad \text{for } y \in [2,3], \\ &= 0 \qquad \text{for } y \notin [1,3], \\ \mu_b(4) &= 1, \qquad \mu_b(5) = \tfrac{1}{2}, \quad \mu_b(z) = 0 \quad \text{for } z \notin \{4,5\}, \end{aligned}$$

can be calculated as follows.

For $x < 4$ the inequality $\frac{x}{z} < 1$ holds for $z = 4$ and $z = 5$ This means that

$$\mu_{a \odot b}(x) = \sup_{z \in R_0} \left(\min \left(\mu_a \left(\frac{x}{z} \right), \mu_b(z) \right) \right) = 0$$

If $x \in [4, 5)$ then $\frac{x}{z} < 1$ for $z = 5$ and

$$\sup_{z \in R_0} \left(\min \left(\mu_a \left(\frac{x}{z} \right), \mu_b(z) \right) \right) = \min \left(\mu_a \left(\frac{x}{4} \right), \mu_b(4) \right) = \mu_a \left(\frac{x}{4} \right) = \frac{x}{4} - 1$$

For $x \in [5,\ 7.5]$

$$\begin{aligned} &\sup_{z \in R_0} \left(\left(\mu_a \left(\frac{x}{z} \right), \mu_b(z) \right) \right) \\ &= \max \left[\min \left(\mu_a \left(\frac{x}{4} \right), \mu_b(4) \right), \right. \\ &\quad \left. \min \left(\mu_a \left(\frac{x}{5} \right), \mu_b(5) \right) \right] \\ &= \max \left[\mu_a \left(\frac{x}{4} \right), \mu_a \left(\frac{x}{5} \right) \right] = \mu_a \left(\frac{x}{4} \right) = \frac{x}{4} - 1 \end{aligned}$$

Analogously for $x \in [7.5,\ 8]$

$$\begin{aligned} &\max \left[\min \left(\mu_a \left(\frac{x}{4} \right), \mu_b(4) \right), \min \left(\mu_a \left(\frac{x}{5} \right) \mu_b(5) \right) \right] \\ &= \max \left[\mu_a \left(\frac{x}{4} \right), \mu_b(5) \right] = \mu_a \left(\frac{x}{4} \right) = \frac{x}{4} - 1 \end{aligned}$$

For $x \in [8,\ 10]$ the procedure is analogous except for another analytic form of $\mu_a\left(\frac{x}{4}\right)$, namely

$$\max\left[\mu_a\left(\frac{x}{4}\right),\ \mu_b(5)\right] = \mu_a\left(\frac{x}{4}\right) = 3 - \frac{x}{4}$$

If $x > 10$ then the maximum value is achieved by $\mu_b(z)$.

For $x \in [10,\ 12]$

$$\max\left[\mu_a\left(\frac{x}{4}\right),\ \mu_b(5)\right] = \mu_b(5) = \frac{1}{2}$$

For $x > 12 \quad \mu_a\left(\frac{x}{4}\right) = 0$ and then

$$\max\left[\min\left(\mu_a\left(\frac{x}{4}\right),\ \mu_b(4)\right),\ \min\left(\mu_a\left(\frac{x}{5}\right),\ \mu_b(5)\right)\right]$$

$$\begin{aligned} = \min\left(\mu_a\left(\frac{x}{5}\right),\ \mu_b(5)\right) &= \mu_b(5) = 0.5 \qquad \text{for } x \in [12, 12.5] \\ &= \mu_a\left(\tfrac{x}{5}\right) = 3 - \tfrac{x}{5} \text{ for } x \in [12.5, 15] \end{aligned}$$

Summarizing the previous results, we obtain

$$\begin{aligned} \mu_{a\odot b}(x) &= 0 && \text{for } x \notin (4, 15), \\ &= \tfrac{x}{4} - 1 && \text{for } x \in [4, 8], \\ &= 3 - \tfrac{x}{4} && \text{for } x \in [8, 10], \\ &= \tfrac{1}{2} && \text{for } x \in [10,\ 12.5], \\ &= 3 - \tfrac{x}{5} && \text{for } x \in [12.5,\ 15] \end{aligned}$$

The membership function $\mu_{a\odot b}$ is evidently continuous, and $\mu_{a\odot b}(4) = \mu_{a\odot b}(15) = 0,\ \mu_{a\odot b}(8) = 1,\ \mu_{a\odot b}(10) = \mu_{a\odot b}(12.5) = 0.5$.

The continuous fuzzy quantities can be multiplied in the following way.

Example 12. Let a, $b \in \mathbb{R}$, analogously to Example 6, such that

$$\begin{aligned} \mu_a(y) &= y - 1 && \text{for } y \in [1, 2], & \mu_b(z) &= 4 - z && \text{for } z \in [3, 4], \\ &= 0 && \text{for } y \notin [1, 2], & &= 0 && \text{for } z \notin [3, 4]. \end{aligned}$$

Then $\mu_{a\odot b}(x) = 0$ for $x \notin (3, 8)$, since for any such x and any $y \in R_0$ at least one of the terms $\mu_a(y)$, $\mu_b\left(\frac{x}{y}\right)$ in (3.7) vanishes.

If $x \in [3, 6]$ then

$$\sup_{z \in R_0}\left(\min\left(\mu_a\left(\frac{x}{z}\right),\ \mu_b(z)\right)\right) = \min\left(\mu_a\left(\frac{x}{3}\right),\ \mu_b(3)\right)$$

$$= \mu_a\left(\frac{x}{3}\right) = \frac{x}{3} - 1$$

If $x \in [6, 8]$ then

$$\sup_{y \in R_0} \left(\min \left(\mu_a(y),\, \mu_b \left(\frac{x}{y} \right) \right) \right) = \min \left(\mu_a(2),\, \mu_b \left(\frac{x}{2} \right) \right)$$

$$= \mu_b \left(\frac{x}{2} \right) = 4 - \frac{x}{2}$$

Hence

$$\begin{aligned} \mu_{a \odot b}(x) &= \tfrac{x}{3} - 1 && \text{for } x \in [3, 6], \\ &= 4 - \tfrac{x}{2} && \text{for } x \in [6, 8], \\ &= 0 && \text{for } x \notin [3, 8]. \end{aligned}$$

The continuity is also preserved in the following product of fuzzy quantities. It is useful to note that the specific symmetry of the input fuzzy quantities preserved by the addition operation is broken by the multiplication. The product of fuzzy quantities processed in Example 7, calculated below, illustrates these facts.

Example 13. Let us consider $a,\, b \in \mathbb{R}$ such that

$$\begin{aligned} \mu_a(y) &= y - 1 && \text{for } y \in [1, 2], & \mu_b(z) &= 5 - z && \text{for } z \in [4, 5], \\ &= 3 - y && \text{for } y \in [2, 3], & &= z - 5 && \text{for } z \in [5, 6], \\ &= 0 && \text{for } y \notin (1, 3), & &= 0 && \text{for } z \notin (4, 6) \end{aligned}$$

Then $\mu_{a \odot b}(x) = 0$ for $x \notin [4, 18]$, since for such x and for any $y \in R_0$ at least one of the values $\mu_a(y),\, \mu_b \left(\frac{x}{y} \right)$ is equal to 0.

Further, if $x \in [4, 8]$ then

$$\sup_{z \in R_0} \left(\min \left(\mu_a \left(\frac{x}{z} \right),\, \mu_b(z) \right) \right) = \max_{z \in [4,5]} \left(\min \left(\mu_a \left(\frac{x}{z} \right),\, \mu_b(z) \right) \right)$$

$$= \min \left(\mu_a \left(\frac{x}{4} \right),\, \mu_b(4) \right) = \mu_a \left(\frac{x}{4} \right) = \frac{x}{4} - 1$$

Analogously, for $x \in [12, 18]$

$$\sup_{z \in R_0} \left(\min \left(\mu_a \left(\frac{x}{z} \right),\, \mu_b(z) \right) \right) = \max_{z \in [5,6]} \left(\min \left(\mu_a \left(\frac{x}{z} \right),\, \mu_b(z) \right) \right)$$

$$= \min \left(\mu_a \left(\frac{x}{6} \right),\, \mu_b(6) \right) = \mu_a \left(\frac{x}{6} \right) = 3 - \frac{x}{6}$$

If $x \in [8, 12]$ then

$$\sup_{y \in R_0} \left(\min \left(\mu_a(y),\, \mu_b \left(\frac{x}{y} \right) \right) \right)$$

$$= \max\left[\sup_{y\in[1,2]}\left(\min\left(\mu_a(y),\, \mu_b\left(\frac{x}{y}\right)\right)\right),\right.$$

$$\left.\sup_{y\in[2,3]}\left(\min\left(\mu_a(y),\, \mu_b\left(\frac{x}{y}\right)\right)\right)\right]$$

where

$$\sup_{y\in[1,2]}\left(\min\left(\mu_a(y),\, \mu_b\left(\frac{x}{y}\right)\right)\right) = \max_{y\in[1,2]}\left(\left(y-1,\, \frac{x}{y}-5\right)\right)$$

and

$$\sup_{y\in[2,3]}\left(\min\left(\mu_a(y),\, \mu_b\left(\frac{x}{y}\right)\right)\right) = \max_{y\in[1,2]}\left(\min\left(3-y,\, 5-\frac{x}{y}\right)\right)$$

Both maxima presented above are achieved for y from the interval under consideration, for which both terms in the minimum expressions are equal. In the first case

$$y-1 = \frac{x}{y}-5 \quad \text{for } y\in[1,2], \quad \text{i.e.,} \quad y = -2+\sqrt{4+x}$$

and in the second case

$$3-y = 5-\frac{x}{y} \quad \text{for } y\in[2,3], \quad \text{i.e.,} \quad y = 1+\sqrt{1+x}$$

Substituted in $\mu_a(y)$ or $\mu_b\left(\frac{x}{y}\right)$, this means that

$$\sup_{y\in[1,2]}\left(\min\left(\mu_a(y),\, \mu_b\left(\frac{x}{y}\right)\right)\right) = -3+\sqrt{4+x}$$

$$\sup_{y\in[2,3]}\left(\min\left(\mu_a(y),\, \mu_b\left(\frac{x}{y}\right)\right)\right) = 4-\sqrt{1+x}$$

The maximum of both expressions is

$$4-\sqrt{1+x} \quad \text{for } x\in[8, x_0]$$

and

$$3+\sqrt{4+x} \quad \text{for } x\in[x_0, 12]$$

where $x_0 \in [8, 12]$ is the value of x for which

$$4-\sqrt{1+x} = -3+\sqrt{4+x}, \quad \text{i.e.,} \quad x_0 \doteq 9.8$$

Summarizing the results of the computations, we can see that

$$
\begin{aligned}
\mu_{a\odot b}(x) &= 0 && \text{for } x \notin (4,18),\\
&= \tfrac{x}{4}-1 && \text{for } x \in [4,8],\\
&= 4-\sqrt{1+x} && \text{for } x \in [8,\ 9.8],\\
&= -3+\sqrt{4+x} && \text{for } x \in [9.8,\ 12],\\
&= 3-\tfrac{x}{6} && \text{for } x \in [12,18].
\end{aligned}
$$

As was mentioned above, the membership function $\mu_{a\odot b}$ is continuous but not symmetric.

Contrary to the previous example, the analytic form of the product of fuzzy quantities given in Example 8 is quite simple.

Example 14. Let us consider $a,\ b \in \mathbb{R}$ such that

$$
\begin{aligned}
\mu_a(y) &= y-1 && \text{for } y \in [1,2], & \mu_b(z) &= 5-z && \text{for } z \in [4,5],\\
&= 1 && \text{for } y \in [2,3], & &= z-5 && \text{for } z \in [5,6],\\
&= 0 && \text{for } y \notin [1,3), & &= 0 && \text{for } z \notin [4,6].
\end{aligned}
$$

Evidently, in this case also $\mu_{a\odot b}(x) = 0$ for $x \notin [4,18)$, since for these x and for any $y \in R_0$ at least one of the values $\mu_a(y)$, $\mu_b\left(\frac{x}{y}\right)$ is equal to 0.

If $x \in [4,8]$ then, analogously to the previous example,

$$\mu_{a\odot b}(x) = \frac{x}{4} - 1$$

If $x \in [8,12]$ then

$$\sup_{z\in R_0}\left(\min\left(\mu_a\left(\frac{x}{z}\right),\ \mu_b(z)\right)\right) = \min\left(\mu_a\left(\frac{x}{4}\right),\ \mu_b(4)\right) = \mu_a\left(\frac{x}{4}\right) = 1$$

For $x \in [12,18]$

$$\sup_{z\in R_0}\left(\min\left(\mu_a\left(\frac{x}{z}\right),\ \mu_b(z)\right)\right) = \min\left(\mu_a\left(\frac{x}{6}\right),\ \mu_b(6)\right) = \mu_a\left(\frac{x}{6}\right) = 1$$

This means that

$$
\begin{aligned}
\mu_{a\odot b}(x) &= 0 && \text{for } x \le 4 \text{ or } x \ge 18,\\
&= \tfrac{x}{4}-1 && \text{for } x \in [4,8],\\
&= 1 && \text{for } x \in [8,18).
\end{aligned}
$$

Multiplication (3.7) is commutative and associative, i.e., for any $a, b, c \in \mathbb{R}$

$$a \odot b = b \odot a \tag{3.11}$$

$$a \odot (b \odot c) = (a \odot b) \odot c \tag{3.12}$$

where the proofs are analogous to those of (3.4) and (3.5); moreover

$$a \odot \langle 1 \rangle = a \tag{3.13}$$

This means that $\mathbb{R}$ with the operation $\odot$ is a *commutative multiplicative monoid.*

The remaining group property, namely the existence of the inverse elements, entails even more difficulties than does its analogy in the additive case. First, the natural inverse $(1/a)$ given by (2.7) makes sense only for $a \in \mathbb{R}_0$. This is analogous to the crisp case, where $1/x$ makes sense only for $x \in R_0$. Much more serious is the fact that there is no general equality between $a \odot (1/a)$ and $\langle 1 \rangle$, as shown by the following simple example. So, neither $\mathbb{R}$ nor $\mathbb{R}_0$ can be a group for the operation $\odot$.

Example 15. If $a \in \mathbb{R}$, $\mu_a(1) = 0.5$, $\mu_a(2) = 1.0$, $\mu_a(x) = 0.0$ for other $x \in R$ then $1/a \in \mathbb{R}^+$, $\mu_{1/a}(1) = 0.5$, $\mu_{1/a}(0.5) = 1.0$, $\mu_{1/a}(x) = 0.0$ for other $x \in R$, and

$$\begin{aligned}
\mu_{a\odot(1/a)}(x) &= 0.5 \quad \text{for } x = 0.5, \\
&= 1.0 \quad \text{for } x = 1, \\
&= 0.5 \quad \text{for } x = 2, \\
&= 0.0 \quad \text{for other } x \in R.
\end{aligned}$$

Having introduced the sum of fuzzy quantities in (3.1), the crisp product in (3.9), and the fuzzy product in (3.7), it is natural to be interested in their distributivity. The results obtained are not always encouraging. If $a, b \in \mathbb{R}$ and $r \in R$ then, by [67] and [68]

$$r \cdot (a \oplus b) = (r \cdot a) \oplus (r \cdot b) \tag{3.14}$$

since for any $x \in R$, $r \neq 0$,

$$\begin{aligned}
\mu_{r\cdot(a\oplus b)}(x) &= \mu_{a\oplus b}(x/r) = \sup_{y\in R} (\min(\mu_a(y), \mu_b(x/r - y)) \\
&= \sup_{z\in R} (\min(\mu_a(z/r), \mu_b((x - z)/r))) \\
&= \sup_{z\in R} (\min(\mu_{r\cdot a}(z), \mu_{r\cdot b}(x - z))) = \mu_{r\cdot a\oplus r\cdot b}(x)
\end{aligned}$$

and, evidently, for $r = 0$

$$\mu_{r\cdot(a\oplus b)}(x) = \mu_{\langle 0\rangle}(x) = \mu_{r\cdot a\oplus r\cdot b}(x)$$

This distributivity offers a useful tool for calculation with fuzzy quantities. Unfortunately, the complementary distributivity rule, namely the equality between $(r_1+r_2)\cdot a$ and $(r_1\cdot a)\oplus(r_2\cdot a)$, for $r_1,\ r_2 \in R,\ a \in \mathbb{R}$, is not generally fulfilled. This is illustrated by the following example.

Example 16. Let $r_1 = r_2 = 1,\ \mu_a(1) = 0.5,\ \mu_a(2) = 1,\ \mu_a(x) = 0$ for $x \notin \{1,2\}$. Then

$$\mu_{a\oplus a}(2) = 0.5, \quad \mu_{a\oplus a}(3) = 0.5, \quad \mu_{a\oplus a}(4) = 1, \quad \mu_{a\oplus a}(x) = 0$$

for other x, and

$$\mu_{2\cdot a}(2) = 0.5, \quad \mu_{2\cdot a}(4) = 1, \quad \mu_{2\cdot a}(x) = 0 \quad \text{for } x \notin \{2,4\}$$

This means that $a \oplus a \neq 2\cdot a$.

As shown above, even very simple distributive equalities, self-evident with crisp quantities, are not guaranteed if the fuzzy case is considered.

The distributivity of the crisp product is, according to (3.9), a special case of the distributivity of $\oplus$ and $\odot$. The assumptions under which $a\odot(b\oplus c)$ is equal to $(a\odot b)\oplus(a\odot c)$ for $a,\ b,\ c \in \mathbb{R}$ have been presented elsewhere (e.g., in [23]). The equality is true iff at least one of the following conditions is fulfilled:

- a is a real number, i.e., $a = \langle y\rangle$ for some $y \in R$; this case is described in (3.14)
- $\mu_a,\ \mu_b,\ \mu_c$ are upper semicontinuous and either $b \in \mathbb{R}^+,\ c \in \mathbb{R}^+$ or $b \in \mathbb{R}^-,\ c \in \mathbb{R}^-$
- $\mu_a,\ \mu_b,\ \mu_c$ are upper semicontinuous and $b = -b$ and $c = -c$

3.3 Exponentiation

The representation principle can be used even when dealing with exponents. To avoid formal complications connected with negative roots and/or exponents, we limit our considerations here to positive fuzzy quantities from $\mathbb{R}^+$ and the signed ones from $\mathbb{R}^*$.

Equation (2.4) tells us that a crisp exponent over a fuzzy root is a special case of the general case where both exponent and root are fuzzy. Nevertheless, to stress some special properties valid for the crisp exponent in a stronger form, we mention both cases separately.

Let $a \in \mathbb{R}^+$ and $r \in R_0$ then the rth *power of* a, denoted by a^r, is a fuzzy quantity with membership function μ_{a^r} such that

$$\begin{aligned} \mu_{a^r}(x) &= \mu_a(x^{1/r}) \quad \text{for all } x > 0 \\ &= 0 \qquad\qquad\; \text{for all } x \leq 0 \end{aligned} \tag{3.15}$$

The second part of (3.15) eliminates the formal problems with $x^{1/r}$ when $r \in R$ and $x < 0$. In accordance with the positivity assumption for a, the definitive setting of $\mu_{a^r}(x) = 0$ for $x \leq 0$ is a formal simplification only.

The concept of the rth power can be illustrated by the following example.

Example 17. Let $r = \frac{1}{2}, a \in \mathbb{R}^+$, where

$$\begin{aligned} \mu_a(x) &= \tfrac{1}{3}x - \tfrac{1}{3} \quad \text{for } x \in [1,4], \\ &= \tfrac{9}{5} - \tfrac{1}{5}x \quad \text{for } x \in [4,9], \\ &= 0 \qquad\qquad \text{for } x \notin [1,9]. \end{aligned}$$

Then (3.15) immediately implies that for $a^r = a^{1/2}$

$$\begin{aligned} \mu_{a^{1/2}}(x) &= \tfrac{1}{3}\sqrt{x} - \tfrac{1}{3} \quad \text{for } x \in [1,2], \\ &= \tfrac{9}{5} - \tfrac{1}{5}\sqrt{x} \quad \text{for } x \in [2,3], \\ &= 0 \qquad\qquad\quad \text{for } x \notin [1,3]. \end{aligned}$$

It is not difficult to prove, as shown in [67] and [23], that for any $a \in \mathbb{R}^+,\ b \in \mathbb{R}^+$,
$r \in R_0,\ s \in R_0$

$$a^{-r} = 1/a^r = (1/a)^r \tag{3.16}$$

which also implies $(a^r)^{-1} = a^{-r}$,

$$(a^r)^s = a^{r \cdot s} \tag{3.17}$$

$$a^r \odot b^r = (a \odot b)^r \tag{3.18}$$

On the other hand, as the inverse elements do not exist (see Section 3.2), the relation (3.16) shows that the product $a^r \odot a^{-r}$ is not generally equal to $\langle 1 \rangle$.

The equality between $a^r \odot a^s$ and a^{r+s} for $a \in \mathbb{R}^+,\ r,\ s \in R_0$, is also not generally fulfilled, as shown by the following example.

Example 18. Let us consider $a \in \mathbb{R}^+$ such that

$$\mu_a(1) = \mu_a(2) = 1, \quad \mu_a(x) = 0 \quad \text{for } 1 \neq x \neq 2.$$

Then

$$\begin{aligned} \mu_{a \odot a}(x) &= 1 \quad \text{for } x = 1, 2, 4, \\ &= 0 \quad \text{for other } x, \end{aligned}$$

and

$$\begin{aligned} \mu_{a^2}(x) &= 1 \quad \text{for } x = 1, 4, \\ &= 0 \quad \text{for other } x. \end{aligned}$$

Hence, $\mu_{a^2}(2) \neq \mu_{a \odot a}(2)$.

Let us concentrate now on the general case of a fuzzy exponent over a fuzzy root. Let $a \in \mathbb{R}^+$ and $b \in \mathbb{R}^*$ be fuzzy quantities. Then the bth *power of* a, denoted by a^b, is a fuzzy quantity with membership function

$$\begin{aligned} \mu_{a^b}(x) &= \sup_{y \neq 0} \left(\min(\mu_a(x^{1/y}),\, \mu_b(y)) \right), \quad \text{for } x > 0 \qquad (3.19) \\ &= 0 \quad \text{for } x \leq 0 \end{aligned}$$

Any crisp number $r \in R_0$ can be considered for the degenerated fuzzy quantity $\langle r \rangle \in \mathbb{R}^*$. Substituting $\langle r \rangle$ for b in (3.19) we obtain definition (3.15), which means that a^r is a special case of a^b for $b = \langle r \rangle$. This fact means that the consequences of Example 18 remain valid even for the fuzzy power a^b, and $a^b \odot a^c$ is not generally equal to $a^{b \oplus c}$ for $a \in \mathbb{R}^+$, $b, c \in \mathbb{R}^*$.

The principal methods of calculation of fuzzy powers can be illustrated by the following examples. In the first, we consider root and exponent with discrete and finite supports.

Example 19. Let $a, b \in \mathbb{R}^+$ be fuzzy quantities with

$$\mu_a(1) = \tfrac{1}{2}, \quad \mu_a(2) = 1, \quad \mu_a(x) = 0 \quad \text{for } x \notin \{1, 2\},$$
$$\mu_b(2) = \tfrac{2}{3}, \quad \mu_b(3) = 1, \quad \mu_b(x) = 0 \quad \text{for } x \notin \{2, 3\}.$$

Then, using (3.19), we can see that $\mu_{a^b}(x) = 0$ for $x \notin \{1, 4, 8\}$, since for such x and for any $y \in R_0$ at least one of the values $\mu_a(x^{1/y})$ and $\mu_b(y)$ in (3.19) vanishes. Further,

$$\begin{aligned} \mu_{a^b}(1) &= \max\left[\min(\mu_a(1),\, \mu_b(2)),\, \min(\mu_a(1),\, \mu_b(3))\right] \\ &= \max\left[\min\left(\frac{1}{2}, \frac{2}{3}\right), \min\left(\frac{1}{2}, 1\right)\right] = \frac{1}{2} \end{aligned}$$

$$\mu_{a^b}(4) = \min\left(\mu_a(2),\, \mu_b(2)\right) = \min\left(1,\, \frac{2}{3}\right) = \frac{2}{3}$$

$$\mu_{a^b}(8) = \min\left(\mu_a(2),\, \mu_b(3)\right) = 1$$

The fuzzy expression with continuous root and discrete exponent, presented in the next example, generalizes the previous example.

Example 20. Let $a,\, b \in \mathbb{R}^+$ be fuzzy quantities such that

$$\begin{aligned} \mu_a(y) &= \tfrac{1}{2}y \quad \text{for } y \in [1,2], \\ &= 0 \quad\ \ \text{for } y \notin [1,2], \end{aligned}$$

(note that $\mu_a(1) = \frac{1}{2}$, $\mu_a(2) = 1$, as in Example 19).

$$\mu_b(2) = \frac{2}{3}, \quad \mu_b(3) = 1, \quad \mu_b(z) = 0 \quad \text{for } z \notin \{2,3\}$$

(also as in Example 19).

Then (3.19) simply implies that $\mu_{a^b}(x) = 0$ for $x \notin [1,8]$, since for any $y \in R_0$ and any such x at least one of the values $\mu_a(x^{1/z})$ and $\mu_b(z)$ is equal to 0.

Further, for $x \in [1,4]$

$$\mu_{a^b}(x) = \max_{z \in R_0}\left[\min(\mu_a(x^{1/2}),\, \mu_b(2)),\, \min(\mu_a(x^{1/3}),\, \mu_b(3))\right]$$

Here

$$\begin{aligned} \min\left(\mu_a(x^{1/2}),\, \mu_b(2)\right) &= \min\left(\frac{x^{1/2}}{2},\, \frac{2}{3}\right) \\ &= \tfrac{1}{2}x^{1/2} \text{ for } x \in \left[1,\, \tfrac{16}{9}\right] \\ &= \tfrac{2}{3} \qquad\ \text{for } x \in \left[\tfrac{16}{9}, 4\right] \end{aligned}$$

$$\min\left(\mu_a(x^{1/3}),\, \mu_b(3)\right) = \mu_a(x^{1/3}) = \frac{1}{2}x^{1/3}$$

Then

$$\begin{aligned} \mu_{a^b}(x) &= \max\left(\min\left(\frac{1}{2}x^{1/2},\, \frac{2}{3}\right),\, \frac{1}{2}x^{1/3}\right) \\ &= \tfrac{1}{2}x^{1/2} \text{ for } x \in \left[1,\, \tfrac{16}{9}\right] \doteq [1,\, 1.78] \\ &= \tfrac{2}{3} \qquad\ \text{for } x \in \left[\tfrac{16}{9},\, \tfrac{64}{27}\right] \doteq [1.78,\, 2.37] \\ &= \tfrac{1}{2}x^{1/3} \text{ for } x \in \left[\tfrac{64}{27},\, 4\right] \doteq [2.37,\, 4] \end{aligned}$$

If $x \in [4, 8]$ then

$$\mu_{a^b}(x) = \min\left(\mu_a(x^{1/3}),\, \mu_b(3)\right) = \mu_a\left(x^{1/3}\right) = \frac{1}{2}x^{1/3}$$

The next example represents further generalization of the fuzzy quantities under consideration to the continuous ones.

Example 21. Let us consider fuzzy quantities $a,\, b \in \mathbb{R}$ such that

$$\begin{aligned} \mu_a(y) &= \tfrac{1}{2}y \quad \text{for } y \in [1, 2] \\ &= 0 \quad\ \ \text{for } y \notin [1, 2] \end{aligned}$$

(as in the previous example),

$$\begin{aligned} \mu_b(z) &= \tfrac{1}{3}z \quad \text{for } z \in [2, 3] \\ &= 0 \quad\ \ \text{for } z \notin [2, 3] \end{aligned}$$

(note that $\mu_b(2) = \frac{2}{3}$, $\mu_b(3) = 1$, as in Examples 19 and 20).

Expression (3.19), defining the value of $\mu_{a^b}(x)$, can be calculated using the following auxiliary results. It is easy to verify that for every $y \in [1, 4]$ and every $z \in (2, 3]$

$$\mu_a(y^{1/2}) > \mu_a(y^{1/z})$$

This means that the same inequality is valid for $z \in \left(2,\, \frac{16}{9}\right]$. Further, for every $z \in [2, 3]$

$$\begin{aligned} \min\left(\mu_a(y^{1/z}),\, \mu_b(z)\right) &= \mu_a(y^{1/z}) = \tfrac{1}{2}y^{1/z} && \text{for } y \in [1, z_y] \\ &= \mu_b(z) = \tfrac{1}{3}z && \text{for } y \in [z_y, 2^z] \\ &= 0 && \text{for } y \notin [1, 2^z] \end{aligned}$$

where z_y is the solution of the equality

$$\mu_a(y^{1/z}) = \mu_b(z)$$

which means

$$\frac{1}{2}y^{1/z} = \frac{1}{3}z$$

Solving this relation, we compute

$$y_z = \left(\frac{2}{3}z\right)^z, \quad \text{and} \quad \mu_a(y_z^{1/z}) = \mu_b(z) = \frac{1}{3}z$$

These results imply the expressions for μ_{a^b} in the form

$$\begin{aligned}\mu_{a^b} &= \tfrac{1}{2}x^{1/2} && \text{for } x \in \left[1,\ \tfrac{16}{9}\right] \\ &= \mu_b(z(x)) = \tfrac{1}{3}z(x) && \text{for } x \in \left[\tfrac{16}{9}, 8\right]\end{aligned}$$

where $z(x)$ as function of x is the solution of the equality with parameter x

$$x^{1/z} = \frac{2}{3}z$$

To illustrate, if $x = \frac{16}{9}$ then $z(x) = 2$ and $\mu_{a^b}(x) = \mu_b(z(x)) = \frac{2}{3}$. Analogously, $z(x) = 2.5$ for $x = \left(\frac{5}{3}\right)^{5/2} = 3\,125\,/\,243 = 3.586$; for this $z(x)$ $\mu_{a^b}(x) = \frac{5}{6}$.

The following table presents more values of $z(x)$ and $\mu_b(z(x)) = \mu_{a^b}(x)$ for x over the interval $\left[\frac{16}{9},\ 8\right]$.

x	$\frac{16}{9} \doteq 1.78$	2.32	3.09	3.586	4.18	5.74	6.76	7.35	8.0
$z(x)$	2	2.2	2.4	2.5	2.6	2.8	2.9	2.95	3.0
$\mu_b(z(x))$	$\frac{2}{3} \doteq 0.67$	0.73	0.8	$\frac{5}{6} = 0.83$	0.87	0.93	0.97	0.98	1.0

Finally, it can easily be verified that, analogously to the previous example,

$$\mu_{a^b}(x) = 0 \quad \text{for } x \notin [1, 8]$$

since for such x and for all $z \in R_0$ either $\mu_a(x^{1/z})$ or $\mu_b(z)$ is equal to 0.

The properties of the fuzzy exponent can easily be derived, as is shown in [67]. Their similarity with the analogous properties of crisp powers is interesting.

If $a \in \mathbb{R}^+,\ b,\ c \in \mathbb{R}^*$ then

$$a^{-b} = (1/a)^b = 1/a^b \tag{3.20}$$

since for any $x > 0$

$$\begin{aligned}\mu_{a^{-b}}(x) &= \sup_{y\neq 0}\left(\min(\mu_a(x^{1/y}),\ \mu_{-b}(y))\right) \\ &= \sup_{y\neq 0}\left(\min(\mu_a(x^{1/y}),\ \mu_b(-y))\right) \\ &= \sup_{z\neq 0}\left(\min(\mu_a(x^{-1/z}),\ \mu_b(z))\right)\end{aligned}$$

$$= \sup_{z \neq 0} \left(\min(\mu_a((x^{-1})^{1/z}), \mu_b(z)) \right)$$

$$= \mu_{a^b}(x^{-1}) = \mu_{1/a^b}(x)$$

$$= \sup_{z \neq 0} \left(\min(\mu_{1/a}(x^{1/z}), \mu_b(z) \right)$$

$$= \sup_{y \neq 0} \left(\min(\mu_a((x^{1/y})^{-1}), \mu_b(y) \right)$$

$$= \sup_{y \neq 0} \left(\min(\mu_{1/a}(x^{1/y}), \mu_b(y)) \right) = \mu_{(1/a)^b}(x)$$

The following property is also analogous to the properties of a crisp exponent over a fuzzy root, namely

$$(a^b)^c = a^{b \odot c} \tag{3.21}$$

In this case, for any $x > 0$

$$\mu_{(a^b)^c}(x) = \sup_{u \neq 0} \left(\min(\mu_{a^b}(x^{1/u}), \mu_c(u)) \right)$$

$$= \sup_{u \neq 0} \left[\min \left(\sup_{v \neq 0} [\min(\mu_a((x^{1/u})^{1/v}), \mu_b(v))], \mu_c(u) \right) \right]$$

$$= \sup_{u \neq 0} \left[\min \left(\sup_{v \neq 0} [\min(\mu_a(x^{1/(u \cdot v)}), \mu_b(v))], \mu_c(u) \right) \right]$$

$$= \sup_{w \neq 0} \left[\min \left(\sup_{v \neq 0} [\min(\mu_a(x^{1/w}), \mu_b(v))], \mu_c(w/v) \right) \right]$$

$$= \sup_{w \neq 0} \left[\min \left(\mu_a(x^{1/w}), \sup_{v \neq 0} [\min(\mu_b(v), \mu_c(w/v))] \right) \right]$$

$$= \sup_{w \neq 0} \left[\min(\mu_a(x^{1/w}), \mu_{b \odot c}(w)) \right] = \mu_{a^{b \odot c}}(x)$$

and (3.21) holds.

If, on the other hand, $a, b \in \mathbb{R}^+$, $c \in \mathbb{R}^*$ then

$$a^c \odot b^c = (a \odot b)^c \tag{3.22}$$

since for any $x > 0$

$$\mu_{a^c \odot b^c}(x)$$

$$= \sup_{y \neq 0} (\min(\mu_{a^c}(x/y), \mu_{b^c}(y)))$$

$$= \sup_{y \neq 0} \left(\min \left[\sup_{z \neq 0} (\min(\mu_a(x^{1/z}/y^{1/z}), \mu_c(z))), \right. \right.$$

$$\sup_{z\neq 0}(\min(\mu_b(y^{1/z}), \mu_c(z)))\Big]\Big)$$

$$= \sup_{y\neq 0}\left[\sup_{z\neq 0}\left(\min[\mu_a(x^{1/z}\,/\,y^{1/z}), \mu_b(y^{1/z}),\, \mu_c(z)]\right)\right]$$

$$= \sup_{z\neq 0}\left[\sup_{y\neq 0}\left(\min[\mu_a(x^{1/z}\,/\,y^{1/z}), \mu_b(y^{1/z}),\, \mu_c(z)]\right)\right]$$

$$= \sup_{z\neq 0}\left(\min\left[\mu_c(z),\, \sup_{y\neq 0}(\min(\mu_a(x^{1/z}\,/\,y^{1/z}),\, \mu_b(y^{1/z})))\right]\right)$$

$$= \sup_{z\neq 0}\left(\min(\mu_c(z),\, \mu_{a\odot b}(x^{1/z}))\right) = \mu_{(a\odot b)^c}(x)$$

as is shown in [67].

It can easily be seen that $a^b \odot a^{-b}$, $a \in \mathbb{R}^+$, $b \in \mathbb{R}^*$ is not generally equal to $\langle 1\rangle$. Example 5 illustrates this fact for $b = \langle 1\rangle$ with regard to (3.16).

It is also natural to consider the symmetric case to a^r, namely r^a for $r \in R, r > 0$, $a \in \mathbb{R}^*$. The defining relation (3.19) immediately implies by substitution of $\langle r\rangle$ for r that (according to [67])

$$\mu_{\langle r\rangle^a}(x) = \mu_a\left(\frac{\ln x}{\ln r}\right) \qquad \text{for all } x > 0 \tag{3.23}$$

and, of course, the general results derived for a^b are valid for $\langle r\rangle^b$, as well.

3.4 Problems and Their Roots

The previous sections of this chapter summarize the existing similarities between fuzzy and crisp quantities concerning their algebraic structure. We have also seen that the mutual correspondence is not complete. The differences can be divided into two principal sorts.

The first type of problems are those of the incompleteness of the group properties, namely the general absence of opposite or inverse elements in the additive or multiplicative cases, respectively. Also the fact that $a^r \odot a^{-r}$, $a \in \mathbb{R}^+$, $r \in R_0$, is not equal to $\langle 1\rangle$ and, more generally, $a^b \odot a^{-b}$, $a \in \mathbb{R}^+$, $b \in \mathbb{R}^*$, is not equal to $\langle 1\rangle$, as mentioned in the preceding section, is of the same nature. In fact, it would be absurd to expect that arithmetic manipulation with vague fuzzy numbers will result in a crisp number, either $\langle 0\rangle$ or $\langle 1\rangle$, for addition or multiplication.

Addition of somehow opposite fuzzy quantities can lead only to some fuzzy zero or, identically, to some fuzzy quantity which is in certain sense equivalent to $\langle 0 \rangle$. Analogously, multiplication of one fuzzy quantity by another in some sense inverse to it cannot result in a crisp $\langle 1 \rangle$ but rather in some fuzzy unit or in a quantity which is somehow equivalent to $\langle 1 \rangle$. The same is true concerning exponential expressions.

Knowing the essence of these problems, we can try to specify the concepts of "fuzzy zero" and "fuzzy unit", and/or the concepts of corresponding equivalences between fuzzy quantities describing the similarity between crisp and fuzzy zeros or units. The majority of the next chapter is devoted to this topic.

The second type of difficulties are those of the absence of distributivity in certain cases. The unpleasant properties of crisp and fuzzy exponents illustrated by Example 7 are also of this kind. They are much more essential and practically unavoidable. Their existence probably follows from the principal difference between certainty and uncertainty. Some properties of crisp numbers simply cannot be transformed into the universe of vague values. Even if every child knows that in the deterministic case $x + x = 2 \cdot x$, the nature of n-times repeated addition and of the nth product is essentially different. They represent two different operations in linear spaces, which can give the same result in the deterministic case, but whose dissimilarity becomes evident if we use fuzzy instead of crisp values.

We will see that the methods successfully used to avoid the first type of problems completely (or almost completely) fail in the case of the second type. These difficulties inevitably involve the proper essence of vagueness.

Part II

Algebra of Fuzzy Quantities

4

Equivalences

As suggested in the previous discussion, some difficulties involving fuzzy quantities can be avoided by using a more realistic *similarity relation* between such quantities, instead of the too demanding *strict equality.* This method is especially effective if the missing group property is to be completed at least in a weaker form.

4.1 Symmetry

Since it is desirable to construct some form of opposite elements in the additive group, the concept of symmetry appears to be the crucial one.

Let $y \in R$ be a real number. A fuzzy quantity $a \in \mathbb{R}$ is called *y-symmetric* iff for any $x \in R$

$$\mu_a(y + x) = \mu_a(y - x) \tag{4.1}$$

The set of all y-symmetric fuzzy quantities is denoted by $\mathbb{S}_y$, and the union

$$\mathbb{S} = \bigcup_{y \in R} \mathbb{S}_y \tag{4.2}$$

is the set of all *symmetric* fuzzy quantities.

The elementary properties of the symmetry are quite evident. Let us mention the most useful ones. It can easily be seen (e.g., [68] and [75]) that any y-symmetric fuzzy quantity $a \in \mathbb{S}_y$ can be decomposed into the sum

$$a = \langle y \rangle \oplus s, \quad \text{where } s \in \mathbb{S}_0 \tag{4.3}$$

Further, if $x, y \in R$, $r \in R$, $a \in \mathbb{S}_x$, $b \in \mathbb{S}_y$, then $a \oplus b \in \mathbb{S}_{x+y}$, $r \cdot a \in \mathbb{S}_{r \cdot x}$, $-a \in \mathbb{S}_{-x}$. This means that the set $\mathbb{S}$ is closed regarding the operations of addition and crisp product.

On the other hand, the fuzzy product $a \odot b$ for $a, b \in \mathbb{S}$ does not generally preserve the symmetry, as shown in the following example for the discrete case, and in Example 13 for the continuous case (the preservation of symmetry by the addition operation is illustrated by Example 7, in which the sum of 2-symmetric and 5-symmetric fuzzy quantities is a 7-symmetric one).

Example 22. Let $a \in \mathbb{S}_2,\ b \in \mathbb{S}_4,$

$$\mu_a(1) = \mu_a(3) = 0.5, \quad \mu_a(2) = 1, \quad \mu_a(x) = 0 \quad \text{for } x \notin \{1,2,3\}$$
$$\mu_b(3) = \mu_b(5) = 1, \quad \mu_b(4) = 0.6, \quad \mu_b(x) = 0 \quad \text{for } x \notin \{3,4,5\}$$

then

$$\mu_{a\odot b}(3) = 0.5, \quad \mu_{a\odot b}(4) = 0.5, \quad \mu_{a\odot b}(5) = 0.5,$$
$$\mu_{a\odot b}(6) = 1, \quad \mu_{a\odot b}(8) = 0.6, \quad \mu_{a\odot b}(9) = 0.5,$$
$$\mu_{a\odot b}(10) = 1, \quad \mu_{a\odot b}(12) = 0.5, \quad \mu_{a\odot b}(15) = 0.5.$$

We can easily see that $a \odot b$ does not belong to $\mathbb{S}_y$ for any $y \in R$.

It is also useful to note that for any $a \in \mathbb{R}$ and $s \in \mathbb{S}_0$, the product $a \odot s$ is also 0-symmetric, since for any $x \in R,\ x \neq 0$

$$\begin{aligned}\mu_{a\odot s}(x) &= \sup_{y\neq 0}\left(\min(\mu_a(y),\, \mu_s(x/y))\right)\\ &= \sup_{y\neq 0}\left(\min(\mu_a(y),\, \mu_s(-x/y))\right) = \mu_{a\odot s}(-x)\end{aligned}$$

and the value $\mu_{a\odot s}(0)$ does not change the fact of 0-symmetry. In this sense the 0-symmetric fuzzy quantities could be considered as a kind of "fuzzy zeros", as mentioned in Section 3.4.

This conclusion is supported also by the fact that for any $a \in \mathbb{R}$ the sum $a \oplus (-a)$ is 0-symmetric, as for any $x \in R$

$$\begin{aligned}\mu_{a\oplus(-a)}(x) &= \sup_{y\in R}\left(\min(\mu_a(y),\, \mu_{-a}(x-y))\right)\\ &= \sup_{y\in R}\left(\min(\mu_{-a}(-y),\, \mu_a(-x+y))\right)\\ &= \sup_{z\in R}\left(\min(\mu_{-a}(z),\, \mu_a(-x-z))\right)\\ &= \mu_{a\oplus(-a)}(-x)\end{aligned}$$

and also $\langle 0\rangle \in \mathbb{S}_0$, which implies that the 0-symmetry possesses all the usual attributes of "zeros".

4.2 Additive Equivalence

The concept of symmetric, especially 0-symmetric, fuzzy quantities can be used to suggest a similarity relation weaker than the strict equality.

We say that fuzzy quantities a, $b \in \mathbb{R}$ are *additively equivalent* and write $a \sim_{\oplus} b$ iff there exist 0-symmetric fuzzy quantities s_1, $s_2 \in \mathbb{S}_0$ such that

$$a \oplus s_1 = b \oplus s_2 \tag{4.4}$$

This symmetric definition allows the derivation of a few sufficient conditions for its validity, which are shown in [64].

Obviously, $a = b$, as well as $a \oplus s = b$ for some $s \in \mathbb{S}_0$, implies $a \sim_{\oplus} b$.

Moreover, if $a \oplus (-b) \in \mathbb{S}_0$ then $a \sim_{\oplus} b$.

If $a \in \mathbb{S}_x$, $b \in \mathbb{S}_y$ for x, $y \in R$ then $a \sim_{\oplus} b$ iff $x = y$. This means that $a \in \mathbb{S}_x$ implies $a \sim_{\oplus} \langle x \rangle$.

The above result in combination with the fact that $a \odot s \in \mathbb{S}_0$ and $a \oplus (-a) \in \mathbb{S}_0$, for any $a \in \mathbb{R}$, $s \in \mathbb{S}_0$ (both mentioned in the previous section) means that for any a, $b \in \mathbb{R}$ $a \odot (b \oplus (-b)) \in \mathbb{S}_0$ and, on the other hand, $(a \odot b) \oplus (a \odot (-b)) = (a \odot b) \oplus (-(a \odot b)) \in \mathbb{S}_0$. Consequently, in this special case the following weakened distributivity based on the equivalence relation holds:

$$a \odot (b \oplus (-b)) \sim_{\oplus} (a \odot b) \oplus (a \odot (-b)) \tag{4.5}$$

It is also easily provable [64] that for any a, b, $c \in \mathbb{R}$

$$a \sim_{\oplus} b \Longleftrightarrow a \oplus c \sim_{\oplus} b \oplus c \tag{4.6}$$

The relation $\sim_{\oplus}$, defined by (4.4), really is an equivalence relation over $\mathbb{R}$. It means that it is *reflexive*, *symmetric*, and *transitive*; symbolically, for any a, b, $c \in \mathbb{R}$

$$a \sim_{\oplus} a \tag{4.7}$$

$$a \sim_{\oplus} b \Longleftrightarrow b \sim_{\oplus} a \tag{4.8}$$

$$a \sim_{\oplus} b \wedge b \sim_{\oplus} c \Longrightarrow a \sim_{\oplus} c \tag{4.9}$$

All these properties are natural consequences of (4.4), where (4.6) can also be used to derive (4.9). The equivalence relation $\sim_{\oplus}$ partitions any set of fuzzy quantities into disjoint equivalence classes. In the following considerations, it is useful to note that the set $\mathbb{S}$ of all symmetric fuzzy quantities is divided into equivalence classes $\mathbb{S}_x$, $x \in R$. The set of all fuzzy quantities $\mathbb{R}$ is also divided into equivalence classes, where the sets $\mathbb{S}_x$ are subsets of some of those classes (and, of course, $\mathbb{S}_x$ and $\mathbb{S}_y$ for $x \neq y$ cannot be subsets of the same equivalence class). The hypothesis that the sets $\mathbb{S}_x$ form separate equivalence classes even in $\mathbb{R}$, not being completed by any non-symmetric elements, was proved only for fuzzy quantities with finite support sets (2.3) in [64] and [68]. The more general validity of this conclusion, however expectable it is, remains an unsolved problem.

4.3 Transversibility

The symmetry and additive equivalence concepts presented above serve as effective tools for the harmonization of the mutual relation between the essential vagueness of fuzzy quantities and the demands of the last group axiom — the existence of opposite elements.

Obviously, these two concepts cannot be effectively used simultaneously even in the multiplicative case. Among other reasons, symmetry is not preserved by multiplication (as illustrated by Example 22), which implies consequences for the validity of desired equivalences between relevant products.

It also means that for the multiplication of fuzzy quantities we must find an adequate type of balance, analogous to that of additive symmetry. We do so in this section, using some concepts suggested in [67] and [68].

We have already seen in Section 3.2 that more complex processing of multiplication over fuzzy quantities should be limited to signed fuzzy quantities from $\mathbb{R}^*$. Respecting this conclusion, we limit the considerations presented below to the set $\mathbb{R}^*$, as well. In this and especially in the next section we will see that the arguments for this limitation are wider and even more fundamental.

If $y \in R_0$ and $a \in \mathbb{R}^*$ then we say that a is *y-transversible* iff for any $x > 0$

$$\mu_a(y \cdot x) = \mu_a(y/x) \tag{4.10}$$

and

$$\mu_a(y \cdot x) = \mu_a(y/x) = 0$$

for $x \leq 0$. The set of all y-transversible fuzzy quantities is denoted by $\mathbb{T}_y$, and the union

$$\mathbb{T} = \bigcup_{y \in R_0} \mathbb{T}_y \tag{4.11}$$

is the set of all *transversible* fuzzy quantities. It is not difficult to derive the following properties of $\mathbb{T}_y$, presented in [67] and [68].

Obviously $\langle y\rangle \in \mathbb{T}_y$ for any $y \in R_0$, and relation (4.10), together with the assumption that we deal with fuzzy quantities from $\mathbb{R}^* = \mathbb{R}^+ \cup \mathbb{R}^-$ exclusively, implies that $\mathbb{T}_y \subset \mathbb{R}^+$ for $y > 0$, and $\mathbb{T}_y \subset \mathbb{R}^-$ for $y < 0$.

If $a \in \mathbb{R}^*$ and $y \in R_0$, then $a \in \mathbb{T}_y$ iff there exists $t \in \mathbb{T}_1$ such that

$$a = \langle y\rangle \odot t$$

If $x, y \in R_0$, $a \in \mathbb{T}_x$, $b \in \mathbb{T}_y$, $r \in R_0$, then $a \odot b \in \mathbb{T}_{x\cdot y}$, $r \cdot a \in \mathbb{T}_{r\cdot x}$, $(1/a) \in \mathbb{T}_{1/x}$, and $-a \in \mathbb{T}_{-x}$.

On the other hand, if a, $b \in \mathbb{T}$ then generally $a \oplus b$ is not transversible. This is illustrated by the following example.

Example 23. Let us consider positive fuzzy quantities a, $b \in \mathbb{R}^+$, such that $a \in \mathbb{T}_2$, $b \in \mathbb{T}_1$, where

$$\mu_a(1) = \mu_a(4) = 1/2, \qquad \mu_a(2) = 1.0, \quad \mu_a(x) = 0 \qquad \text{for other } x,$$
$$\mu_b(1/2) = \mu_b(2) = 1.0, \quad \mu_b(x) = 0 \qquad \text{for other } x.$$

Then we can easily verify that

$$\mu_{a\oplus b}(1.5) = \mu_{a\oplus b}(3) = \mu_{a\oplus b}(4.5) = \mu_{a\oplus b}(6) = 1/2,$$
$$\mu_{a\oplus b}(2) = \mu_{a\oplus b}(4) = 1.0, \qquad \mu_{a\oplus b}(x) = 0 \qquad \text{for other } x,$$

and $a \oplus b$ is neither transversible nor symmetric. As an illustration, if we compute $a \odot b$, then

$$\mu_{a\odot b}(0.5) = \mu_{a\odot b}(8) = 1/2,$$
$$\mu_{a\odot b}(1) = \mu_{a\odot b}(2) = \mu_{a\odot b}(4) = 1, \qquad \mu_{a\odot b}(x) = 0 \quad \text{for other } x,$$

and, consequently, $a \odot b \in \mathbb{T}_2$.

Further examples of transversible fuzzy quantities are presented in Section 8; in particular, Examples 33 and 34 are quite illustrative.

4.4 Multiplicative Equivalence

Analogously to the additive case, the concept of transversibility can also be used to define a specific type of equivalence. This is based on the idea that in multiplication (3.6), 1-transversible fuzzy quantities play the role of the "fuzzy unit", analogous to the "fuzzy zeros" from $\mathbb{S}_0$ in the additive case.

If a, $b \in \mathbb{R}^*$ are signed fuzzy quantities, then we say that they are *multiplicatively equivalent* and write $a \sim_\odot b$ iff there exist 1-transversible t_1, $t_2 \in \mathbb{T}_1$ such that

$$a \odot t_1 = b \odot t_2 \tag{4.12}$$

It is not difficult to verify (see [67]) that relation $\sim_\odot$ is really an equivalence over $\mathbb{R}^*$, meaning that it is reflexive, symmetric, and transitive. Symbolically, for a, b, $c \in \mathbb{R}^*$

$$\begin{gathered} a \sim_\odot a, \quad a \sim_\odot b \Longleftrightarrow b \sim_\odot a \\ a \sim_\odot b \ \wedge \ b \sim_\odot c \Longrightarrow a \sim_\odot c \end{gathered} \tag{4.13}$$

The following properties of the multiplicative equivalence, described in [67] and [68], illustrate the analogy with its additive counterpart. For a, b, $c \in \mathbb{R}^*$

$$a \sim_\odot b \Longleftrightarrow a \odot c \sim_\odot b \odot c. \tag{4.14}$$

If $t \in \mathbb{T}_y$, then evidently $t \sim_\odot \langle y \rangle$, and if $a \in \mathbb{T}_x$, $b \in \mathbb{T}_y$, then $a \sim_\odot b$ iff $x = y$.

It is also easily verifiable that $a \odot (1/a) \in \mathbb{T}_1$ for $a \in \mathbb{R}^*$; moreover, if $a \odot (1/b) \in \mathbb{T}_1$, then $a \sim_\odot b$ for a, $b \in \mathbb{R}^*$.

The equivalence relation $\sim_\odot$ partitions the sets $\mathbb{R}^*$ and $\mathbb{T}$ into disjoint equivalence classes. Set $\mathbb{T}$ is partitioned into equivalence classes $\mathbb{T}_y$, $y \in R_0$. In the case of the $\mathbb{R}^*$ the structure of equivalence classes is richer. If the support sets (2.3) are finite, then by [79] the sets $\mathbb{T}_y$ are equivalence classes even in $\mathbb{R}^*$. The more general validity of this result has not been verified.

The limitation of the transversibility concept and equivalence relation to signed fuzzy quantities is rational. If $a \in \mathbb{R}_0 - \mathbb{R}^*$, then $a \odot (1/a) \in \mathbb{R}_0 - \mathbb{R}^*$ too; moreover,

$$\mu_{a\odot(1/a)}(x) = \mu_{a\odot(1/a)}(1/x) \quad \text{for } x \in R_0$$

This means that the product $a \odot (1/a)$ has some properties of 1-transversible fuzzy quantities, but it does not belong to $\mathbb{T}_1 \subset \mathbb{R}^*$. There also do not exist t_1, $t_2 \in \mathbb{T}_1$ such that

$$a \odot (1/a) \odot t_1 = \langle 1 \rangle \odot t_2$$

This discrepancy is a serious obstacle to the introduction of more general (applicable to $\mathbb{R}_0$) concepts of transversibility and multiplicative equivalence, specifically the construction of some kind of "fuzzy unit" from $\mathbb{R}_0$ equivalent to $\langle 1 \rangle$, with all its important properties.

Even a modification of transversibility, including non-signed fuzzy quantities leads to serious difficulties. Let us a consider a fuzzy quantity

$a \in \mathbb{R}$ such that

$$\mu_a(y \cdot x) = \mu_a(y/x) \quad \text{for all } x \in R_0, \tag{4.15}$$

where $y \in R_0$ (compare with Equation 4.10), and let us denote the set of such fuzzy quantities by $\mathbb{T}^*_y$. Then evidently

$$\langle -1 \rangle \in \mathbb{T}^*_1, \quad \langle 1 \rangle \in \mathbb{T}^*_{-1}, \quad \mathbb{T}^*_y \cap \mathbb{S}_0 \neq \emptyset, \quad \mathbb{T}^*_1 \cap \mathbb{T}^*_{-1} \neq \emptyset, \quad \text{etc.}$$

This means that $\mathbb{T}^*_1$ cannot be used in the definition of the multiplicative equivalence. Evidently $a \odot (1/a) \in \mathbb{T}^*_1$ for any $a \in \mathbb{R}_0$, but also $a \odot (1/a) \in \mathbb{T}^*_{-1}$, $\langle y \rangle$ would be equivalent with $\langle -y \rangle$, $y \in R_0$. If we denote by $\langle -1, 1 \rangle \in \mathbb{R}_0 - \mathbb{R}^*$ the 0-symmetrical fuzzy quantity that

$$\begin{aligned} \mu_{\langle -1,1 \rangle}(x) &= 1 \quad \text{if } |x| = 1 \\ &= 0 \quad \text{for other } x \in R \end{aligned}$$

then $\langle -1, 1 \rangle \in \mathbb{T}^*_1 \cap \mathbb{T}^*_{-1}$, and $a \odot \langle -1, 1 \rangle \in \mathbb{S}_0$ as mentioned in Section 4.1. Also $(-a) \odot \langle -1, 1 \rangle \in \mathbb{T}^* \cap \mathbb{T}^*_{-1}$ and

$$a \odot \langle -1, 1 \rangle = (-a) \odot \langle -1, 1 \rangle$$

for any $a \in \mathbb{R}_0$. This also means that $a \sim_{\odot} (-a)$. These paradoxical consequences of the proposed modification of the transversibility concept make it unacceptable for our purposes — to be applicable for guaranteeing usual, not controversial, algebraic properties of the operation of multiplication $\odot$. On the contrary, the properties of $\mathbb{T}^*_y$ mentioned above could make the structure of $\mathbb{R}_0$ regarding $\odot$ and the equivalence based on $\mathbb{T}^*_1$ in (4.12) critically strange.

4.5 Uniform Equivalences

Despite the fact that the additive and multiplicative relations of equivalence offer useful properties for relevant operations, they are incompatible with each other. We can expect that the single situation in which they are true simultaneously is the case where $a = b$, $a, b \in \mathbb{R}^*$, when $a \sim_{\oplus} b$ and $a \sim_{\odot} b$.

It is possible to look for some uniform equivalence covering both, $\sim_{\oplus}$ and $\sim_{\odot}$, as its special cases, and preserving their useful properties. This topic was mentioned, for example, in [68]. It might be reasonable to construct the uniform relation as some kind of combination of both special ones. The simplest combinations, namely the logical conjunction or disjunction of the additive and multiplicative equivalence, are not satisfactory. If we say that $a, b \in \mathbb{R}^*$ are uniformly equivalent iff $a \sim_{\oplus} b$ or $a \sim_{\odot} b$ then the resulting relation is not transitive. If, on the other hand, a and b are said to be uniformly equivalent iff $a \sim_{\oplus} b$ and $a \sim_{\odot} b$ then the equivalence classes are too narrow (evidently one-element) to be useful for the purposes for which the equivalences were introduced.

This means that the combination of both special equivalences should be rather more complex. Due to [68] we say that $a,\, b \in \mathbb{R}^*$ are *weakly equivalent* and write $a \approx b$ iff there exist a sequence of fuzzy quantities $c_1, \ldots, c_n \in \mathbb{R}^*$ such that $a = c_1,\ b = c_n$, and

$$c_i \sim_{\oplus} c_{i+1} \quad \text{or} \quad c_i \sim_{\odot} c_{i+1} \quad \text{for all } i = 1, \ldots, n-1 \tag{4.16}$$

It can easily be verified that $\approx$ is an equivalence relation, reflexive, symmetric, and transitive. Of course $a \sim_{\oplus} b$ as well as $a \sim_{\odot} b$ imply $a \approx b$ for $a,\, b \in \mathbb{R}^*$.

If $r \in R_0,\ a,\, b \in \mathbb{R}^*$ then $a \approx b$ iff $r \cdot a \approx r \cdot b$. If, moreover, $t \in \mathbb{T}_1,\ s \in \mathbb{S}_0$, then $a \approx b$ iff $t \odot (a \oplus s) \approx t \odot (b \oplus s)$, under the assumption that $a \oplus s$ and $b \oplus s$ remain signed, and $a \approx b$ if $(a \odot t) \oplus s \approx (b \odot t) \oplus s$ (even in this case, under the assumption that the resulting fuzzy quantities remain signed). By the way, as $a \sim_{\odot} t \odot a$ and $a \sim_{\oplus} a \oplus s$, the equivalence $a \oplus s \approx t \odot a$ is evident.

The weak equivalence also preserves a certain degree of distributivity. Namely, as shown in [68], for $a \in \mathbb{R}^*,\ s \in \mathbb{S}_0,\ t \in \mathbb{T}_1$

$$t \odot (a \oplus s) \approx (t \odot a) \oplus (t \odot s)$$

as

$$t \odot (a \oplus s) \sim_{\odot} a \oplus s \sim_{\oplus} a \sim_{\odot} t \odot a \sim_{\oplus} (t \odot a) \oplus s' \sim_{\oplus} (t \odot a) \oplus (t \odot s)$$

where $t \odot s = s' \in \mathbb{S}_0$.

In spite of these properties the concept of weak equivalence (and its eventual modifications) does not seem to be useful for effective processing of fuzzy quantities in real applications. The equivalence classes of $\approx$ are too large for reasonable distinctions among different types of fuzziness; the formal structure of this weak equivalence is also rather too complicated.

The endeavor to find some uniform equivalence, similarly useful for addition and multiplication, is probably hopeless.

5

Weak Algebraic Properties

The main reasons for defining symmetry and transversibility, and the consequent definitions of additive and multiplicative equivalences, are to guarantee as well the validity of the classical commutative group properties for the addition $\oplus$ on $\mathbb{R}$ and multiplication $\odot$ on $\mathbb{R}^*$, respectively. An immediate consequence of the previous chapter is that this goal is achievable, and the desired group properties are valid in a weaker form, with respect to the specificity of "fuzzy zero" or "fuzzy unit".

5.1 Additive Group Properties

Considering the additive equivalence instead of the strict equality offers some interesting possibilities. First, it is useful to remember that for $a, b \in \mathbb{R}$ the equality $a = b$ implies $a \sim_{\oplus} b$ (see Section 4.2). This means that equalities (3.4), (3.5), and (3.6) are also fulfilled as equivalences

$$a \oplus b \sim_{\oplus} b \oplus a \tag{5.1}$$

$$a \oplus (b \oplus c) \sim_{\oplus} (a \oplus b) \oplus c \tag{5.2}$$

$$a \oplus \langle 0 \rangle \sim_{\oplus} a \tag{5.3}$$

for $a, b, c \in \mathbb{R}$.

Moreover, relation (5.3) can be extended for $a \in \mathbb{R},\ s \in \mathbb{S}_0$ in the form

$$a \oplus s \sim_{\oplus} a \tag{5.4}$$

which corresponds with the idea of 0-symmetry as a fuzzy zero.

Finally, for any $a \in \mathbb{R}$, as mentioned in Section 4.2,

$$a \oplus (-a) \in \mathbb{S}_0, \qquad \text{i.e.} \qquad a \oplus (-a) \sim_{\oplus} \langle 0 \rangle \tag{5.5}$$

and also $a \oplus (-a) \sim_{\oplus} s$ for any $s \in \mathbb{S}_0$, as evidently follows from the definition of symmetry.

This means that any set of representatives of the disjoint equivalence classes into which the set $\mathbb{R}$ is partitioned forms an additive commutative group with regard to the equivalence relation $\sim_\oplus$, with $(-a)$ as an opposite element to $a \in \mathbb{R}$ and with $\mathbb{S}_0$ as a class of fuzzy zeros. In this sense $\mathbb{R}$ is a group with regard to $\sim_\oplus$ and with the characteristics mentioned in the previous paragraph.

The fuzzy zero property of elements from $\mathbb{S}_0$ is completed by a result mentioned in Section 4.1, namely by the fact that $a \odot s \in \mathbb{S}_0$ for any $a \in \mathbb{R},\ s \in \mathbb{S}_0$.

5.2 Linearity of Symmetric Quantities

The symmetry concept and additive equivalence of the fuzzy quantities relation allow the formulation of another result concerning distributivity of the crisp product. Let $r_1, r_2 \in R,\ a \in \mathbb{S}$ be symmetric. Then

$$(r_1 + r_2) \cdot a \sim_\oplus (r_1 \cdot a) \oplus (r_2 \cdot a) \tag{5.6}$$

as follows from (4.3) and (3.14), for $a \in \mathbb{S}_y,\ y \in R$ (see [75]), in the following way.

$$\begin{aligned}(r_1 + r_2) \cdot a &= (r_1 + r_2) \cdot (\langle y \rangle \oplus s)\\ &= (r_1 + r_2) \cdot \langle y \rangle \oplus (r_1 + r_2) \cdot s \sim_\oplus (r_1 + r_2) \cdot y\\ &= r_1 \cdot y + r_2 \cdot y\\ &= r_1 \cdot \langle y \rangle \oplus r_2 \cdot \langle y \rangle \sim_\oplus r_1 \cdot \langle y \rangle \oplus r_1 \cdot s \oplus r_2 \cdot \langle y \rangle \oplus r_2 \cdot s\\ &= r_1(\langle y \rangle \oplus s) \oplus r_2(\langle y \rangle \oplus s) = r_1 \cdot a \oplus r_2 \cdot a\end{aligned}$$

where $a = \langle y \rangle \oplus s,\ s \in \mathbb{S}_0,\ y \in R$. We have used the fact that, due to (2.4), $y \cdot a = \langle y \rangle \odot a$ and by (3.1) $y + a = \langle y \rangle \oplus a$ for any $a \in \mathbb{R},\ y \in R$, where by $y + a$ we denote the fuzzy quantity with the membership function given in (3.3)

$$\mu_{y+a}(x) = \mu_a(x - y) \quad \text{for any } x \in R$$

This result, together with (3.14) implying

$$r \cdot (a \oplus b) \sim_\oplus r \cdot a \oplus r \cdot b, \quad r \in R,\ a, b \in \mathbb{R} \tag{5.7}$$

and together with (5.1), (5.2), (5.3), and (5.5), means that the set $\mathbb{S}$ of symmetric fuzzy quantities is a linear space where the additive equivalence is considered instead of the strict equality.

The previous conclusions are not as effective as they might seem at first sight. The additive equivalence is based on ignoring the symmetric differences between fuzzy quantities. Consequently, those differences between some fuzzy quantities and crisp numbers cannot influence the validity of the classical properties of numbers for symmetric fuzzy quantities — with respect to the equivalence relation. At any rate, the algebraic linearity of $\mathbb{S}$ means that using linear operations — addition and crisp product — we cannot change the principal symmetry of the input quantities. This can be useful if, for example, real data with symmetric fuzzy noise are processed.

5.3 Multiplicative Group Properties

Analogously to the additive case, the multiplicative equivalence can be used to complete the group properties in a weakened form for the multiplication operation. Due to Sections 4.3 and 4.4 we limit our following considerations to the signed fuzzy quantities from $\mathbb{R}^* = \mathbb{R}^+ \cup \mathbb{R}^-$. Even in this case the equality $a = b$, a, $b \in \mathbb{R}^*$, implies $a \sim_{\odot} b$. Thanks to this fact relations (3.8), (3.9), and (3.10) imply, for a, b, $c \in \mathbb{R}^*$, that

$$a \odot b \sim_{\odot} b \odot a \tag{5.8}$$

$$a \odot (b \odot c) \sim_{\odot} (a \odot b) \odot c \tag{5.9}$$

$$a \odot \langle 1 \rangle \sim_{\odot} a \tag{5.10}$$

and also for any $t \in \mathbb{T}_1$, $a \in \mathbb{R}^*$,

$$a \odot t \sim_{\odot} a \tag{5.11}$$

which corresponds to the idea of 1-transversibility as the attribute of the fuzzy unit.

The properties of multiplicative equivalence and transversibility imply for any $a \in \mathbb{R}^*$

$$a \odot (1/a) \in \mathbb{T}_1, \quad \text{i.e.} \quad a \odot (1/a) \sim_{\odot} \langle 1 \rangle \tag{5.12}$$

and also $a \odot (1/a) \sim_{\odot} t$ for any $t \in \mathbb{T}_1$, as mentioned in Section 4.4.

These results mean that the partition of $\mathbb{R}^*$ into multiplicative equivalence $\sim_{\odot}$ classes defines a family of disjoint subsets of $\mathbb{R}^*$ whose arbitrary representatives form a commutative multiplicative group with the inverse elements $1/a$ for $a \in \mathbb{R}^*$, with the 1-transversible fuzzy quantities as the unit elements, and with regard to the multiplicative equivalence

$\sim_{\odot}$ instead of the strict equality [68]. In this sense $\mathbb{R}^*$ can be considered a multiplicative (and commutative) group with the characteristics mentioned in the previous paragraph.

It can also be verified that for any $t \in \mathbb{T}_y,\ y \in R_0$, and for $a,\ b \in \mathbb{R}^*$

$$t \odot (a \oplus b) \sim_{\odot} (y \cdot a) \oplus (y \cdot b) = (\langle y \rangle \odot a) \oplus (\langle y \rangle \odot b) \tag{5.13}$$

since, because of Section 4.3,

$$\begin{aligned} t \odot (a \oplus b) &= (\langle y \rangle \odot t') \odot (a \oplus b) \sim_{\odot} \langle y \rangle \odot (a \oplus b) \\ &= y \cdot (a \oplus b) = (y \cdot a) \oplus (y \cdot b) = (\langle y \rangle \odot a) \oplus (\langle y \rangle \odot b) \end{aligned}$$

where $t' \in \mathbb{T}_1,\ t = \langle y \rangle \odot t'$, and the fact that (2.4) implies the equality $\langle y \rangle \odot a = y \cdot a$ for any $a \in \mathbb{R},\ y \in R$. This relation, shown in [68], represents one of the special cases of distributivity.

5.4 Some Results with Exponents

The multiplicative equivalence can also be used to derive a few results, complementing those presented in Section 3.3. With respect to the method presented therein, we focus on positive fuzzy quantities. The following results are derived in [67].

The properties of 1-transversibility, described in the previous chapter, immediately imply for $t \in \mathbb{T}_1$ and $b \in \mathbb{R}_0$

$$t^b \in \mathbb{T}_1, \quad \text{i.e.} \quad t^b \sim_{\odot} t \tag{5.14}$$

since

$$\begin{aligned} \mu_{t^b}(x) &= \sup_{y \neq 0} \left(\min(\mu_t(x^{1/y}),\ \mu_b(y)) \right) \\ &= \sup_{y \neq 0} \left(\min(\mu_t(1/x^{1/y},\ \mu_b(y)) \right) = \mu_{t^b}(1/x) \end{aligned}$$

This also means that $t^r \sim_{\odot} t$ for $t \in \mathbb{T}_1,\ r \in R_0$.

If $y > 0,\ a \in \mathbb{T}_y$ and $r,\ s \in R_0,\ r + s \neq 0$, then

$$a^r \odot a^s \sim_{\odot} a^{r+s} \tag{5.15}$$

since for $t \in \mathbb{T}_1$ such that $a = \langle y \rangle \odot t$

$$\begin{aligned} a^r \odot a^s &= \langle y \rangle^r \odot t^r \odot \langle y \rangle^s \odot t^s \sim_{\odot} \langle y \rangle^r \odot \langle y \rangle^s \\ &= \langle y^{r+s} \rangle \sim_{\odot} \langle y^{r+s} \rangle \odot t^{r+s} = a^{r+s} \end{aligned}$$

If $a \in \mathbb{R}^+$, $b \in \mathbb{R}_0$, then

$$a^b \odot a^{-b} \in \mathbb{T}_1, \quad \text{i. e.} \quad a^b \odot a^{(-b)} \sim_{\odot} \langle 1 \rangle \tag{5.16}$$

On the other hand, some of the classical properties of exponents, valid for crisp numbers, are not fulfilled in the fuzzy case, even if the equivalence is considered. So, e. g., for $y > 0$, $a \in \mathbb{T}_y$ and $t \in \mathbb{T}_1$ the relations

$$a^t \in \mathbb{T}_y \quad \text{or} \quad \langle y \rangle^t \in \mathbb{T}_y$$

are not valid for $y \neq 1$, as shown by the following example.

Example 24. Let $y = 4$, $t \in \mathbb{T}_1$,

$$\mu_t(1/2) = \mu_t(2) = 1, \quad \mu_t(x) = 0 \quad \text{for } 1/2 \neq x \neq 2.$$

Then

$$\mu_{\langle y \rangle^t}(2) = \mu_{\langle y \rangle^t}(16) = 1, \quad \mu_{\langle y \rangle^t}(x) = 0, \quad 2 \neq x \neq 16$$

and evidently $\langle y \rangle^t \notin \mathbb{T}_y$.

It is useful to remember Example 18, where the multiplicative equivalence between $a^r \odot a^s$ and a^{r+s} for $a \in \mathbb{T}_{\sqrt{2}}$, $r = s = 1$ can be seen. Of course, in Example 18 $a \odot a$ as well as a^2 belong to $\mathbb{T}_2$. It is not difficult to derive from the procedure following Equation 5.15, that given the assumptions of (5.15) both $a^r \odot a^s$ and a^{r+s} belong to $\mathbb{T}_2$ where $z = y^{r+s}$.

Part III

Related Topics

6

Multidimensional Case

The concepts and methods used above for one-dimensional fuzzy quantities can be easily transformed to the multi-dimensional case, and some meaningful new results for vector fuzzy quantities can be derived. Here we briefly present some of them.

It is generally possible to distinguish two principal cases. In the first, we consider n-dimensional fuzzy quantities as fuzzy subsets of the n-dimensional real space R^n. In the second, we deal with n-dimensional vectors of particular fuzzy quantities defined over the set R. This distinction can also be expressed notationally. The set of fuzzy subsets of R^n will be denoted by $\mathbb{R}_n$, and the set of n-dimensional vectors of fuzzy quantities from $\mathbb{R}$, i.e., the n-tuple Cartesian product of $\mathbb{R}$ will be denoted by $\mathbb{R}^n$. Obviously, $\mathbb{R}^1 = \mathbb{R}_1 = \mathbb{R}$.

6.1 Fuzzy Quantities over R^n

Since R^n is a linear space with the coordinate-wise operations of addition and multiplication by a scalar number, i.e.,

$$\mathbf{x} + \mathbf{y} = (x_1 + y_1, \ldots, x_n + y_n) \quad \text{and} \quad r \cdot \mathbf{x} = (r \cdot x_1 + \cdots + r \cdot x_n)$$

for $\mathbf{x} = (x_1, \ldots, x_n) \in R^n$, $\mathbf{y} = (y_1, \ldots, y_n) \in R^n$, $r \in R$, many of the results and procedures mentioned above can be easily generalized to the fuzzy quantities defined as fuzzy subsets of R^n.

A fuzzy subset a of R^n is called a *fuzzy vector quantity* iff

$$\sup(\mu_a(\mathbf{x}) : \mathbf{x} \in R^n) = 1 \tag{6.1}$$

and there exists a compact set $U \subset R^n$ such that

$$\mu_a(\mathbf{x}) = 0 \quad \text{for all } \mathbf{x} \in R^n - U \tag{6.2}$$

The set of all fuzzy vector quantities is denoted by $\mathbb{R}_n$. Formulas (3.1) and (3.10) for addition and multiplication by a scalar value can be used without any essential change.

The extension of formula (3.6) for the multiplication of fuzzy vector quantities is not as simple as the previous ones. It is generally applicable if the multiplication by the inverse of vector $\mathbf{y} \in R^n$ in $\mathbf{x}/\mathbf{y} = \mathbf{x} \cdot (1/\mathbf{y})$ (where $\mathbf{x} \in R^n$, $\mathbf{y} \in R^n$, $1/\mathbf{y}$ is the inverse element of $\mathbf{y}$) is correct. The result of the product of fuzzy vector quantities is either a scalar fuzzy quantity or a fuzzy matrix (understood as a fuzzy subset of the set of $n \times n$ real matrices) determined by the orientation of the input vectors.

It is useful for later applications to mention the concept of the *linear transformation* of fuzzy vector quantities. Let

$$\mathbf{T} = (t_{ij})_{i=1,\ldots,n,\, j=1,\ldots,m}, \; t_{ij} \in R \tag{6.3}$$

be an $n \times m$ real matrix of linear transformation of vectors from $\mathbb{R}_n$, and let us denote by $\mathbf{t}_j$ the jth column of $\mathbf{T}$:

$$\overline{\mathbf{t}}_j = (t_{1j}, \ldots, t_{nj}), \quad j = 1, \ldots, m$$

($\overline{\mathbf{t}}_j$ is the transposition of the column $\mathbf{t}_j$). If $a \in \mathbb{R}_n$ is a fuzzy vector quantity then the product $a \cdot \mathbf{t}_j$ is also a fuzzy quantity. Due to the principles of linear algebra, $a \cdot \mathbf{t}_j$ is a scalar fuzzy quantity from $\mathbb{R}$, and its membership function $\mu_{a \cdot \mathbf{t}_j} : R \to [0, 1]$ is given by

$$\mu_{a \cdot \mathbf{t}_j}(z) = \sup \{\mu_a(\mathbf{x}) : \mathbf{x} \in R^n \text{ such that } z = \mathbf{x} \cdot \mathbf{t}_j\}, \quad z \in R \tag{6.4}$$

Formula (6.4) is naturally interpretable in accordance with the general principles mentioned in sections 1.4 and 2.3. This means that the possibility of achieving the value z for the scalar fuzzy quantity $a \cdot \mathbf{t}_j$ is given by the possibility that for at least one $\mathbf{x} \in R^n$ with membership value $\mu_a(\mathbf{x})$ the product $\mathbf{x} \cdot \mathbf{t}_j$ is equal to z. It is easy to verify that $a \cdot \mathbf{t}_j$ really is a fuzzy quantity, i.e., it fulfills (2.1) and (2.2).

Since the product of $a \in \mathbb{R}_n$ and a single column $\mathbf{t}_j$ of the crisp matrix $\mathbf{T}$ is a scalar fuzzy quantity, then the matrix product $a \cdot \mathbf{T}$ results in a vector of fuzzy quantities $(a \cdot \mathbf{t}_1, \ldots, a \cdot \mathbf{t}_n)$. It is worth mentioning that, even if a is a fuzzy vector quantity from $\mathbb{R}_n$, the product is an m-dimensional vector of fuzzy quantities from $\mathbb{R}^m$ if we use the notation suggested in the introductory paragraphs of this chapter.

6.2 Multidimensional Fuzzy Quantities

As mentioned above, it is also possible to consider an n-dimensional vector whose components are fuzzy quantities from $\mathbb{R}$. Such an n-tuple Cartesian product of $\mathbb{R}$ is denoted by $\mathbb{R}^n$. Evidently, this concept of multidimensional fuzzy quantities assumes the independence of particular scalar fuzzy quantities forming the vectors under consideration. In this sense the vectors of this type represent a rather more special type of multidimensional fuzziness than do the previous fuzzy vector quantities.

In this section a *vector of fuzzy quantities* means the n-tuple $\mathbf{a} = (a_1, \ldots, a_n) \in \mathbb{R}^n$. If $\mathbf{a}$, $\mathbf{b} \in \mathbb{R}^n$, $r \in R$, then $\mathbf{a} \oplus \mathbf{b}$ means the coordinatewise addition

$$\mathbf{a} \oplus \mathbf{b} = (a_1 \oplus b_1, \ldots, a_n \oplus b_n) \in \mathbb{R}^n \tag{6.5}$$

and the crisp product $r \cdot \mathbf{a}$ means the coordinatewise product

$$r \cdot \mathbf{a} = (r \cdot a_1, \ldots, r \cdot a_n) \in \mathbb{R}^n \tag{6.6}$$

without any formal difficulties. Also the product $\mathbf{a} \odot \mathbf{b}$ can be defined in the usual sense. For $\mathbf{a}$ (row vector) and $\mathbf{b}$ (column vector), the product

$$\mathbf{a} \odot \mathbf{b} = (a_1 \odot b_1) \oplus \cdots \oplus (a_n \odot b_n) \in \mathbb{R} \tag{6.7}$$

is a scalar fuzzy quantity, and in the opposite case ($\mathbf{a}$ being a column and $\mathbf{b}$ being a row) this product forms a square $n \times n$ matrix of fuzzy quantities $a_i \odot b_j \in \mathbb{R}$, $i, j = 1, \ldots, n$. Since the distributivity of operations over fuzzy quantities is not guaranteed, it is necessary to respect the order of operations in (6.7) and to maintain that order in all computations of this kind.

The simpler formal structure of vectors from $\mathbb{R}^n$ also means a moderate simplification of their linear transformation. Let $\mathbf{T}$ be an $n \times m$ matrix (6.3), and let $\mathbf{a} \in \mathbb{R}^n$ be a vector of fuzzy quantities. Then the linear transformation given by the product $\mathbf{a} \cdot \mathbf{T}$ transforms $\mathbf{a}$ into $\mathbf{b} \in \mathbb{R}^m$ such that for any $j = 1, \ldots, m$

$$b_j = t_{1j} \cdot a_1 \oplus \cdots \oplus t_{nj} \cdot a_n \in \mathbb{R} \tag{6.8}$$

Similarly to the previous case, the lack of distributivity means that it is necessary to respect the convention regarding the order of operations. On the other hand the properties of addition and crisp product summarized in Sections 3.1 and 3.2 imply that for $\mathbf{a} \in \mathbb{R}^n$, $\mathbf{b} \in \mathbb{R}^n$

$$(\mathbf{a} \oplus \mathbf{b}) \cdot \mathbf{T} = \mathbf{a} \cdot \mathbf{T} \oplus \mathbf{b} \cdot \mathbf{T} \tag{6.9}$$

since

$$\begin{aligned}
(\mathbf{a} \oplus \mathbf{b}) \cdot \mathbf{T} &= ((a_1 \oplus b_1) \cdot t_{1j} \oplus \cdots \oplus (a_n \oplus b_n) \cdot t_{nj})_{j=1,\ldots,m} \\
&= ((a_1 \cdot t_{1j}) \oplus (b_1 \cdot t_{1j}) \oplus \cdots \oplus (a_n \cdot t_{nj}) \oplus (b_n \cdot t_{nj}))_{j=1,\ldots,m} \\
&= (a_1 \cdot t_{1j} \oplus \cdots \oplus a_n \cdot t_{nj})_{j=1,\ldots,m} \oplus \\
&\quad\ (b_1 \cdot t_{1j} \oplus \cdots \oplus b_n \cdot t_{nj})_{j=1,\ldots,m} \\
&= a \cdot \mathbf{T} \oplus \mathbf{b} \cdot \mathbf{T}
\end{aligned}$$

where the usual extension of notations from scalar to vector operations is used. These same properties imply the following consequences: if $\mathbf{a}$, $\mathbf{b}$, $\mathbf{c} \in \mathbb{R}^n$ and $r \in R$, then evidently $\mathbb{R}^n$ is an additive commutative monoid with

$$\mathbf{a} \oplus \mathbf{b} = \mathbf{b} \oplus \mathbf{a}, \ \ \mathbf{a} \oplus (\mathbf{b} \oplus \mathbf{c}) = (\mathbf{a} \oplus \mathbf{b}) \oplus \mathbf{c} \text{ and } \mathbf{a} + \langle 0 \rangle^n = \mathbf{a}$$

where $\langle 0 \rangle^n \in R^n$ is an n-dimensional zero-vector. Moreover,

$$r \cdot (\mathbf{a} \oplus \mathbf{b}) = r \cdot \mathbf{a} \oplus r \cdot \mathbf{b}$$

If we denote by $\mathbb{S}_0^n$ the set of n-dimensional vectors of 0-symmetric fuzzy quantities, then, analogously to Section 4.2, $\mathbf{a} \sim_\oplus \mathbf{b}$ iff there exist $\mathbf{s}_1$, $\mathbf{s}_2 \in \mathbb{S}_0^n$ such that

$$\mathbf{a} \oplus \mathbf{s}_1 = \mathbf{b} \oplus \mathbf{s}_2$$

For any $\mathbf{s} \in \mathbb{S}_0^n$

$$\mathbf{a} \oplus \mathbf{s} \sim_\oplus \mathbf{a}$$

and if we denote by $(-\mathbf{a}) \in \mathbb{R}^n$ the vector of opposite elements $(-\mathbf{a}) = ((-a_1), \ldots, (-a_n))$ to the elements of $\mathbf{a}$ (see Section 2.1), then

$$\mathbf{a} \oplus (-\mathbf{a}) \in \mathbb{S}_0^n, \quad \text{i.e., } \mathbf{a} \oplus (-\mathbf{a}) \sim_\oplus \mathbf{s}$$

for any $\mathbf{s} \in \mathbb{S}_0^n$. This means that $\mathbb{R}^n$ is an additive group with respect to the equivalence relation $\sim_\oplus$, where the equivalence classes of $\sim_\oplus$ are considered.

Let us remember here the notation $\mathbb{S} = \bigcup_{y \in R} \mathbb{S}_y$. By $\mathbb{S}^n$ we denote the n-tuple Cartesian product of $\mathbb{S}$, where it is apparent that $\mathbf{s} = (s_1, \ldots, s_n) \in \mathbb{S}^n$ iff there exists an n-dimensional crisp vector $\mathbf{y} = (y_1, \ldots, y_n) \in R^n$ such that $s_i \in \mathbb{S}_{y_i}$ for any $i = 1, \ldots, n$.

The linearity of the set $\mathbb{S}$ with respect to the equivalence $\sim_\oplus$ (the fact that for any $a \in \mathbb{S}$, r_1, $r_2 \in R$ the equivalence $(r_1 + r_2) \cdot a \sim_\oplus r_1 \cdot a \oplus r_2 \cdot a$ holds) implies that $\mathbb{S}^n$ is a linear vector space if the equivalence $\sim_\oplus$ and corresponding equivalence classes are considered instead of the strict equality. This result is not valid for the whole set $\mathbb{R}^n$ nor for any of its other important subsets.

6.3 Linear Combination

Vectors of fuzzy quantities from $\mathbb{R}^n$ can be processed by methods analogous to those developed for crisp vectors from R^n, but only some of the properties derived for the deterministic case remain valid. This general fact concerns the concepts of linear combination and linear dependence.

Let $\mathbf{a}^{(1)}, \dots, \mathbf{a}^{(m)} \in \mathbb{R}^n$ be vectors of fuzzy quantities and $r_1, \dots, r_m \in R$ be real numbers. Then the vector $\mathbf{a} = (a_1, \dots, a_n) \in \mathbb{R}^n$ such that

$$\mathbf{a} = r_1 \cdot \mathbf{a}^{(1)} \oplus \cdots \oplus r_m \cdot \mathbf{a}^{(m)} \tag{6.10}$$

i.e.,

$$a_i = r_1 \cdot a_i^{(1)} \oplus \cdots \oplus r_m \cdot a_i^{(m)}, \quad i = 1, \dots, n$$

is the *linear combination* of vectors $\mathbf{a}^{(1)}, \dots, \mathbf{a}^{(m)}$. Due to the absence of one of the distributivity laws, it is necessary to respect convention with regard to the order of arithmetic operations — we therefore calculate the crisp products before the additions even if the initial (real or fuzzy) quantities are the same.

When considering the specificity of the subtraction of fuzzy quantities we say that the vectors $\mathbf{a}^{(1)}, \dots, \mathbf{a}^{(m)} \in \mathbb{R}^n$ are *linearly dependent* iff there exist real numbers $r_1, \dots, r_m \in R$ such that $r_1^2 + \cdots + r_m^2 > 0$ and the linear combination (6.10) is 0-symmetric, i.e.,

$$r_1 \cdot \mathbf{a}^{(1)} \oplus \cdots \oplus r_m \cdot \mathbf{a}^{(m)} \in \mathbb{S}_0^n \tag{6.11}$$

which means that for any $i = 1, \dots, n$

$$r_1 \cdot a_i^{(1)} \oplus \cdots \oplus r_m \cdot a_i^{(m)} \sim_\oplus \langle 0 \rangle \tag{6.12}$$

This concept of linear dependence is very similar to the deterministic one, and it also preserves some properties of the deterministic version, as shown in [76]. Namely, if vectors $\mathbf{a}^{(1)}, \dots, \mathbf{a}^{(m)} \in \mathbb{R}^n$ are linearly dependent, then for any $\mathbf{a}^{(m+1)} \in \mathbb{R}^n$ the vectors $\mathbf{a}^{(1)}, \dots, \mathbf{a}^{(m)}, \mathbf{a}^{(m+1)}$ are also linearly dependent.

If for some $r_2, \dots, r_m \in R$

$$\mathbf{a}^{(1)} \sim_\oplus r_2 \cdot \mathbf{a}^{(2)} \oplus \cdots \oplus r_m \cdot \mathbf{a}^{(m)}$$

then vectors $\mathbf{a}^{(1)}, \dots, \mathbf{a}^{(m)}$ are linearly dependent and, on the other hand, if vectors $\mathbf{a}^{(1)}, \dots, \mathbf{a}^{(m)}$ are linearly dependent, then for any $k \in \{1, \dots, m\}$ there exist real numbers $r_1, \dots, r_{k-1}, r_{k+1}, \dots, r_m$ such that

$$\mathbf{a}^{(k)} \sim_\oplus r_1 \cdot \mathbf{a}^{(1)} \oplus \cdots \oplus r_{k-1} \cdot \mathbf{a}^{(k-1)} \oplus r_{k+1} \cdot \mathbf{a}^{(k+1)} \oplus \cdots \oplus r_m \cdot \mathbf{a}^{(m)}$$

6.4 Equivalentions

The validity of group properties in the form of equivalences instead of equalities is the reason for the effort to define and study the weakened form of such fundamental algebraic structures as equations. Accomplishing this task, at least on the elementary level, means that more complex processing of fuzzy data, namely the derivation of an unknown fuzzy variable from algebraic "equation-like" structures is feasible. This effort is supported by the opinion that the additive equivalence $\sim_{\oplus}$, ignoring the influence of "fuzzy zeros", can adequately represent the relation of similarity between fuzzy quantities. An attempt to solve "*equivalentions*" (this term denotes equations where the equivalence relation is considered instead of the equality) was made in [80], and the results, even if some of them are negative, offer an elementary view of the problem.

Here we consider the following type of equivalentions — linear ones with fuzzy variables, crisp coefficients, and, consequently, fuzzy right-hand-side. Such equivalentions have the form

$$a_1 \cdot x_1 \oplus \cdots \oplus a_n \cdot x_n \sim_{\oplus} r \tag{6.13}$$

where $r \in \mathbb{R}$, $x_1, \ldots, x_n \in \mathbb{R}$ and $a_1, \ldots, a_n \in R$, $a_i \neq 0$, $i = 1, \ldots, n$.

First, we deal with the simplest special form of (6.13), the equivalention of one variable

$$a \cdot x \oplus (-r) \sim_{\oplus} \langle 0 \rangle \quad \text{or} \quad a \cdot x \sim_{\oplus} r \tag{6.14}$$

with x, $r \in \mathbb{R}$, $a \in R$, $a \neq 0$. The properties of $\sim_{\oplus}$, namely relation (4.6), imply that

$$a \cdot x \oplus (-r) \sim_{\oplus} \langle 0 \rangle \text{ iff } a \cdot x \oplus (-r) \oplus r \sim_{\oplus} r \text{ iff } a \cdot x \oplus s \sim_{\oplus} r \text{ iff } a \cdot x \sim_{\oplus} r$$

for some $s = (-r) \oplus r \in \mathbb{S}_0$. Analogously,

$$a \cdot x \oplus (-r) \sim_{\oplus} b \quad \text{iff} \quad a \cdot x \sim_{\oplus} b \oplus r$$

for any $b \in \mathbb{R}$. If b is crisp, i.e., $b \in R$, then evidently $b \oplus r$ denotes $\langle b \rangle \oplus r$. The solution of (6.14) is evidently a fuzzy quantity, $x \in \mathbb{R}$, fulfilling

$$x \sim_{\oplus} (1/a) \cdot r \tag{6.15}$$

A more general case is the system of n linear equivalentions with n fuzzy variables

$$\begin{array}{c} a_{11} \cdot x_1 \oplus \cdots \oplus a_{1n} \cdot x_n \sim_{\oplus} r_1 \\ \cdots\cdots\cdots\cdots\cdots\cdots\cdots\cdots\cdots \\ a_{n1} \cdot x_1 \oplus \cdots \oplus a_{nn} \cdot x_n \sim_{\oplus} r_n \end{array} \tag{6.16}$$

with $a_{ij} \in R$, $r_i \in \mathbb{R}$, $x_i \in \mathbb{R}$, i, $j = 1, \ldots, n$. In this multi-variable case the classical deterministic algorithms for solving system of equations cannot be mechanically transformed to the equivalentions case. The inherent difficulties are illustrated by the following example.

Example 25. Let us consider 2 equivalentions of 2 variables

$$a_{11} \cdot x_1 \oplus a_{12} \cdot x_2 \sim_\oplus r_1$$
$$a_{21} \cdot x_1 \oplus a_{22} \cdot x_2 \sim_\oplus r_2$$

Then it is correct to calculate

$$a_{21} \cdot x_1 \sim_\oplus r_2 \oplus (-a_{22}) \cdot x_2$$

and

$$x_1 \sim_\oplus (1/a_{21}) \cdot r_2 \oplus (-a_{22}/a_{21}) \cdot x_2 \tag{6.17}$$

and to substitute

$$(a_{11}/a_{21}) \cdot r_2 \oplus (-a_{22} \cdot a_{11}/a_{21}) \cdot x_2 \oplus a_{12} \cdot x_2 \sim_\oplus r_1$$

but the absence of the distributivity law does not allow the extraction of the variable x_2, i.e., to rearrange (6.17) into something like

$$(a_{12} - (a_{11} \cdot a_{22}/a_{21})) \cdot x_2 \sim_\oplus r_1 \oplus (-a_{11}/a_{21}) \cdot r_2.$$

Such a procedure can be accepted only by a convention strictly determining the order of the arithmetic operations used during the solution procedure. Similar difficulties connected with the absence of distributivity may appear even if some other solution methods are used. Nevertheless, some of them assume the fixed ordering of operations, which seemingly compensates for the lack of distributivity. This concerns, for example, the determinant method. In our case we may compute the determinants

$$D = a_{11} \cdot a_{22} - a_{12} \cdot a_{21} \in R \tag{6.18}$$

$$D_1 = (a_{12} \cdot r_2) \oplus ((-a_{22}) \cdot r_1) \in \mathbb{R} \tag{6.19}$$

$$D_2 = (a_{11} \cdot r_2) \oplus ((-a_{21}) \cdot r_1) \in \mathbb{R}$$

and put $x_1 \sim_\oplus (1/D) \cdot D_1$, $x_2 \sim_\oplus (1/D) \cdot D_2$. However, this procedure in the deterministic case is justified by an algebraic proof in which the distributivity is unavoidable. This means that application of the determinant method is incorrect. It can be accepted only by a convention based on the analogy with the deterministic equations, but is not supported by serious theoretical proof.

The discussion presented in the previous example is of general validity. The solution methods analogous to the deterministic ones are, more or less obviously, dependent on the application of the distributivity law and, consequently, they cannot be correctly used to solve systems like

(6.16). As follows from Section 5.2, if the right-hand-side fuzzy quantities are symmetric then the distributivity steps are always correct, and the classical solution methods can be applied. The linearity of $\mathbb{S}$ means that the symmetric right-hand-side implies also the symmetry of the left-hand-side fuzzy quantities, so that the operations being applied are admissible. At any rate, as also mentioned in Section 5.2, the combination of additive equivalence and symmetry represents a trivial approach.

A rather clearer situation takes place in the case of one equivalention with n variables, as in (6.13). Analogously to Section 6.3, we may rearrange (6.13) into

$$x_1 \sim_{\oplus} (1/a_1) \cdot r \oplus (-a_2/a_1) \cdot x_2 \oplus \cdots \oplus (-a_n/a_1) \cdot x_n \tag{6.20}$$

which specifies the relation between x_1 and the linear combination of other fuzzy variables.

Up to this point, we have been interested in equivalentions with a fuzzy right hand side, fuzzy variables, and crisp coefficients. The opposite case (fuzzy coefficients and crisp variables) is rather specific. If a crisp variable can be a result of any equivalention with fuzzy elements, it is additively equivalent to some fuzzy quantity. The definition of the additive equivalence means that the only fuzzy quantities which are equivalent to a crisp number x are x-symmetric fuzzy quantities (see Equations (4.3) and (4.4)). This implies a strong limitation for fuzzy coefficients and the fuzzy right-hand-side of the considered equivalention. Thus symmetry is proved for fuzzy quantities with finite supports ([79] and [64]) and it is highly probable that the others are symmetric too. In consequence, this means that such equivalentions represent only the classical deterministic manipulation with crisp numbers (or degenerated fuzzy quantities) completed by symmetric fuzzy noise, where the existence of the additive equivalence $\sim_{\oplus}$ instead of equality makes this noise practically negligible. Something similar can be seen in Section 5.2 regarding the equivalent distributivity of symmetric fuzzy quantities.

The theoretical tools derived in Chapter 4 can be used to solve simple equivalentions in which all elements — variables, coefficients, and right hand sides — are fuzzy. In such a case the solution method depends on the type of equivalence in the relation. If $a \in \mathbb{R}^*$, $r \in \mathbb{R}$ and $x \in \mathbb{R}$ then

$$a \odot x \sim_{\odot} r \tag{6.21}$$

can be solved by $x \sim_{\odot} r \odot (1/a)$, meanwhile

$$a \odot x \sim_{\oplus} r$$

is not generally manageable.

7

Fuzzy Numbers and Fuzzy Intervals

The model of fuzzy quantities represents a generalization of various special models of vague numeric values. Important and theoretically interesting is the concept of fuzzy numbers (and fuzzy intervals), which, by means of a simple model, covers a wide class of practical applications. Both fuzzy numbers and fuzzy intervals are fuzzy quantities with partwise monotonic membership functions of rather specific simple type (see [23], [24], [35], [52], and [80]).

7.1 Linear Fuzzy Numbers

Following [23] and other papers, we use the term *fuzzy number* for any fuzzy quantity $a \in \mathbb{R}$ (i.e., fulfilling (2.1) and (2.2)) with the following specific properties:

$$\text{There exists exactly one } x_0 \in R \text{ such that } \mu_a(x_0) = 1 \tag{7.1}$$

$$\text{For any } x,\, y \in R,\ x \leq y, \text{ and any } z \in [x, y]\ \mu_a(z) \geq \min(\mu_a(x),\, \mu_a(y)) \tag{7.2}$$

A value x_0 satisfying (7.1) is called the *modal value* of the fuzzy number a. Conditions (7.1) and (7.2) imply that $\mu_a(x)$ is not decreasing over the interval $(-\infty, x_0)$ and not increasing over (x_0, ∞).

An extremely simple type of fuzzy number is represented by a so called *linear fuzzy number* $a \in \mathbb{R}$ (in some papers called a *triangular fuzzy number*), which is fully characterized by three real numbers $a_1 < a_0 < a_2$, $a_1,\, a_0,\, a_2 \in R$, such that

$$\mu_a(a_0) = 1 \tag{7.3}$$

$$\mu_a(x) = 0 \quad \text{for } x \leq a_1 \text{ or } x \geq a_2 \tag{7.4}$$

$$\text{if } x \in [a_1, a_0],\ x = \lambda\, a_0 + (1 - \lambda)\, a_1, \text{ then } \mu_a(x) = \lambda \tag{7.5}$$

$$\text{if } x \in [a_0, a_2],\ x = \lambda\, a_0 + (1-\lambda)\, a_2,\ \text{then } \mu_a(x) = \lambda \tag{7.6}$$

where $\lambda \in [0,1]$ and a_0 is the modal value of a. Let us denote by $\mathbb{L} \subset \mathbb{R}$ the set of all linear fuzzy numbers.

7.2 Arithmetic of Fuzzy Numbers

The standard arithmetic operations defined for fuzzy quantities by (3.1), (3.7), and (3.12) are used even for fuzzy numbers.

If $a \in \mathbb{R}$ is a fuzzy number and $r \in R$, then $r \cdot a$ fulfills (7.1) and (7.2), and is evidently a fuzzy number. In the case of the addition and fuzzy product the preservation of (7.2) represents a rather less evident question, which is presented in detail in [23] and [24], for example. The exact properties of the arithmetic operations over fuzzy numbers depend on the analytic form of their membership functions. Let us examine a few typical cases.

If $a \in \mathbb{R}$ is a fuzzy number with x_0, x_1, $x_2 \in R$ where $\mu_a(x_0) = 1$, $\mu_a(x_1) = \mu_a(x_2) = 0$, $x_1 < x_2$, and if $\mu_a : R \to [0,1]$ is continuous, then (7.1) and (7.2) obviously imply that $\mu_a(x) = 0$ for $x \notin (x_1, x_2)$. Moreover, μ_a is not decreasing over $[x_1, x_0]$ and not increasing over $[x_0, x_2]$, as mentioned above.

Consider fuzzy numbers a, $b \in \mathbb{R}$ with x_0, $y_0 \in R$, where $\mu_a(x_0) = \mu_b(y_0) = 1$. Let $c = a \oplus b$. It is easy to prove that $\mu_c(z_0) = 1$ for $z_0 = x_0 + y_0$ and that $\mu_c(z) < 1$ for any $z \neq z_0$. This means that c fulfills (7.1). Consider z_1, $z_2 \in R$ such that $z_1 < z_2$, and a number $z \in [z_1, z_2]$. Then the weak monotonicity of μ_a, μ_b and the addition formula (3.1) imply that there exist x_1, $x_2 \in R$, and y_1, $y_2 \in R$ such that

$$\mu_c(z_1) = \min\left(\mu_a(x_1),\, \mu_b(y_1)\right), \quad z_1 = x_1 + y_1 \tag{7.7}$$

and, analogously,

$$\mu_c(z_2) = \min\left(\mu_a(x_2),\, \mu_b(y_2)\right), \quad z_2 = x_2 + y_2 \tag{7.8}$$

Then, for any x, $y \in R$ such that $x + y = z$, x is between x_1 and x_2 (i.e., either in $[x_1, x_2]$ or in $[x_2, x_1]$ up to their ordering), and y is between y_1 and y_2, formula (3.1) implies that

$$\mu_c(z) \geq \min\left(\mu_a(x),\, \mu_b(y)\right)$$

The validity of (7.2) for a and b combined with (7.7) and (7.8) obviously implies that

$$\mu_c(z) \geq \min\left(\mu_c(z_1),\, \mu_c(z_2)\right)$$

and (7.2) is fulfilled for c, too.

An analogous procedure can be used if we consider fuzzy numbers with integral values, where (7.1) and (7.2) as well as their consequences are related to the integers x, y, z. Even in this case the sum, but not the crisp product, remains a fuzzy number with integral values.

Even stronger results can be derived for linear fuzzy numbers. Let $a \in \mathbb{L}$ be such a fuzzy number with values a_0, a_1, $a_2 \in R$ fulfilling (7.3), (7.4), (7.5), and (7.6). If $r \in R$ then obviously $r \cdot a$ is a linear fuzzy number from $\mathbb{L}$ with the modal value $r \cdot a_0$ and with $r \cdot a_1$ and $r \cdot a_2$ as the values fulfilling (7.4), (7.5), and (7.6), where the positivity or negativity of r is significant for the ordering of $r \cdot a_1$ and $r \cdot a_2$. If, moreover, $b \in \mathbb{L}$ is a linear fuzzy number with the determining triple b_0, b_1, b_2, then it is possible to see that $a_0 + b_0$ is the unique modal value of $a \oplus b$, and that $\mu_{a\oplus b}(z) = 0$ for $z \notin (a_1 + b_1, a_2 + b_2)$, which means that (7.4) is true. Let us consider

$$z = \lambda(a_0 + b_0) + (1 - \lambda)(a_1 + b_1), \quad \lambda \in [0,1]$$

Then $z = x + y$, where

$$x = \lambda a_0 + (1 - \lambda) a_1, \qquad y = \lambda b_0 + (1 - \lambda) b_1$$

and because of (3.1) $\mu_{a\oplus b}(z) \geq \lambda$, since $\mu_a(x) = \lambda$, $\mu_b(y) = \lambda$. If $\mu_{a\oplus b}(z) > \lambda$, then there exist $x' \in [a_1, a_0]$, $y' \in [b_1, b_0]$ such that $x' + y' = z$ and

$$\min(\mu_a(x', \mu_b(y')) > \lambda$$

This contradicts the fact that either $x' < x$ and consequently $\mu_a(x') < \lambda$ or, vice versa, $y' < y$ and $\mu_b(y') < \lambda$. The last condition (7.6) can be derived for $a \oplus b$ in an analogous way. Consequently, $a \oplus b \in \mathbb{L}$, as well.

This means that the set $\mathbb{L}$ of linear fuzzy numbers is invariant regarding the crisp product and addition of its members. As we already know, each linear fuzzy number is fully determined by three crisp real values, denoted here by (a_1, a_0, a_2). Summarizing the conclusions derived above, we see that for $a \in \mathbb{L}$, $b \in \mathbb{L}$, determined by triples (a_1, a_0, a_2), (b_1, b_0, b_2), respectively, and for $r \in R$ also $r \cdot a \in \mathbb{L}$, $a \oplus b \in \mathbb{L}$, and they are determined by the triples

$$(r \cdot a_1, r \cdot a_0, r \cdot a_2) \text{ for } r \geq 0, \ (r \cdot a_2, r \cdot a_0, r \cdot a_1) \text{ for } r < 0, \tag{7.9}$$

$$(a_1 + b_1, \ a_0 + b_0, \ a_2 + b_2)$$

respectively. In this way the operations over $\mathbb{L}$ are reduced to the operations over triples of crisp real numbers that are fully distributive.

This implies that for $a \in \mathbb{L}$ and r_1, $r_2 \in R$ the fuzzy quantities $(r_1 + r_2) \cdot a$ and $(r_1 \cdot a) \oplus (r_2 \cdot a)$ are also linear fuzzy numbers. It is easy to verify that for $r_1 > 0$, $r_2 < 0$ and for a determined by the triple

(a_1, a_0, a_2), $r_1 \cdot a$ is also determined by $(r_1 \cdot a_1,\ r_1 \cdot a_0,\ r_1 \cdot a_2)$ and $r_2 \cdot a$ is determined by $(r_2 \cdot a_2,\ r_2 \cdot a_0,\ r_2 \cdot a_1)$. Moreover, if $r_1 \geq 0,\ r_2 \geq 0$, then $(r_1 + r_2) \cdot a$ is evidently determined by

$$((r_1 + r_2) \cdot a_1,\ (r_1 + r_2) \cdot a_0,\ (r_1 + r_2) \cdot a_2)$$

$(r_1 \cdot a) \oplus (r_2 \cdot a)$ is determined by

$$\begin{aligned} &(r_1 \cdot a_1 + r_2 \cdot a_1,\ r_1 \cdot a_0 + r_2 \cdot a_0,\ r_1 \cdot a_2 + r_2 \cdot a_2) \\ &= ((r_1 + r_2) \cdot a_1,\ (r_1 + r_2) \cdot a_0,\ (r_1 + r_2) \cdot a_2) \end{aligned}$$

and, consequently,

$$(r_1 + r_2) \cdot a = (r_1 \cdot a) \oplus (r_2 \cdot a)$$

Analogously, if $r_1 \leq 0,\ r_2 \leq 0$, then

$$(r_1 + r_2) \cdot a = (r_1 \cdot a) \oplus (r_2 \cdot a)$$

and this linear fuzzy number is determined by the triple

$$((r_1 + r_2) \cdot a_2,\ (r_1 + r_2) \cdot a_0,\ (r_1 + r_2) \cdot a_1)$$

Using an analogous procedure it is not difficult to verify that for $r_1 > 0,\ r_2 < 0$ generally

$$(r_1 + r_2) \cdot a \oplus s = (r_1 \cdot a) \oplus (r_2 \cdot a)$$

where $s \in \mathbb{L} \cap \mathbb{S}_0$ is such that

$$\begin{aligned} &s = (r_2(a_2 - a_1),\ 0,\ r_2(a_1 - a_2)) \quad \text{if} \ \ |r_1| > |r_2| \\ &s = (r_1(a_1 - a_2),\ 0,\ r_1(a_2 - a_1)) \quad \text{if} \ \ |r_2| > |r_1| \\ &s = (0, 0, 0) \quad \text{if} \ \ |r_1| = |r_2| \end{aligned}$$

This result means that for $r_1 > 0,\ r_2 < 0$

$$(r_1 + r_2) \cdot a \sim_{\oplus} (r_1 \cdot a) \oplus (r_2 \cdot a)$$

and the distributivity is fulfilled at least in the form of equivalence.

The distributivity in the equivalence form implies that it is possible to solve systems of n equivalentions with n variables (introduced in Section 6.4 with Equation 6.16) by means of classical algebraic methods.

If $a \in \mathbb{L}$, then $(-a) \in \mathbb{L}$ is determined by $(-a_2,\ -a_0,\ -a_1)$ if a is determined by (a_1, a_0, a_2). It means that $a \oplus (-a) \in \mathbb{L}$ is a 0-symmetric linear fuzzy number determined by $(a_1 - a_2,\ 0,\ a_2 - a_1)$.

If $a \in \mathbb{L},\ b \in \mathbb{L}$ are linear fuzzy numbers determined by the triples (a_1, a_0, a_2) and (b_0, b_1, b_2), respectively, then we can apply formulas

(2.8), (2.9), and (2.10) and easily derive the following values of possibility of the relations discussed in Chapter 2. The equality $a_0 = b_0$ immediately implies that

$$\mu_{\sim}(a,b) = 1.0, \quad \mu_{\succsim}(a,b) = \mu_{\succsim}(b,a) = 1.0, \quad \mu_{\succ}(a,b) = \mu_{\succ}(b,a) = 1.0 \tag{7.10}$$

If $a_0 < b_0$ and $a_2 > b_1$, then the linearity of the considered fuzzy numbers leads to the following results:

$$\mu_{\sim}(a,b) = \mu_{\sim}(b,a) = \lambda_0 \tag{7.11}$$

where $\lambda_0 \in [0,1]$ is the value of λ for which

$$\lambda_0\, a_0 + (1-\lambda_0)\cdot a_2 = \lambda_0\, b_0 + (1-\lambda_0)\cdot b_1$$

i.e.,

$$\lambda_0 = \frac{b_1 - a_2}{a_0 - b_0 + b_1 - a_2} \tag{7.12}$$

Similarly

$$\mu_{\succ}(b,a) = \mu_{\succsim}(b,a) = 1.0 \tag{7.13}$$

and

$$\mu_{\succ}(a,b) = \mu_{\succsim}(b,a) = \lambda_0 \tag{7.14}$$

for λ_0 given by (7.12).

If not only $a_0 < b_0$ but also $a_2 \le b_1$, then evidently

$$\mu_{\sim}(a,b) = \mu_{\sim}(b,a) = 0$$

and

$$\mu_{\succ}(b,a) = \mu_{\succsim}(b,a) = 1.0$$
$$\mu_{\succ}(a,b) = \mu_{\succsim}(a,b) = 0.0$$

7.3 Fuzzy Interval

A fuzzy interval is a fuzzy quantity generalizing the concept of a fuzzy number as follows: a fuzzy quantity $a \in \mathbb{R}$ is a *fuzzy interval* if

$$\text{there exist } x_0,\ x_0' \in R,\ x_0 < x_0', \text{ such that} \tag{7.15}$$
$$\mu_a(x) = 1 \text{ for } x \in [x_0, x_0'],\ \mu_a(x) < 1 \text{ for } x \notin [x_0, x_0']$$

and (7.2) is fulfilled. Values $x \in [x_0, x']$ are called the *modal values* of the fuzzy interval a.

Analogously to linear fuzzy numbers we can define the linear fuzzy interval $a \in \mathbb{R}$, also called a *trapezoidal fuzzy number*, as a fuzzy quantity fully determined by the quadruple (a_1, a_0, a_0', a_2) of real numbers such that $a_1 < a_0 < a_0' < a_2$ and

$$\mu_a(x) = 1 \text{ iff } x \in [a_0, a_0'] \tag{7.16}$$

$$\mu_a(x) = 0 \text{ iff } x \notin [a_1, a_2] \tag{7.17}$$

$$\mu_a(x) = \lambda \text{ for } x = \lambda\, a_0 + (1-\lambda)\, a_1 \tag{7.18}$$

$$\mu_a(x) = \lambda \text{ for } x = \lambda\, a_0 + (1-\lambda)\, a_2 \tag{7.19}$$

where $\lambda \in [0,1]$. Analogously to the previous section, it is possible to show that for fuzzy intervals a, b, and real $r \in R$, $r \cdot a$ and $a \oplus b$ are also fuzzy intervals. If a, b are linear then $r \cdot a$ and $a \oplus b$ are also linear, and they are determined by quadruples

$$\begin{aligned}
&(r \cdot a_1,\ r \cdot a_0,\ r \cdot a_0',\ r \cdot a_2) \quad \text{if } r \geq 0 \qquad (7.20)\\
&(r \cdot a_2,\ r \cdot a_0',\ r \cdot a_0,\ r \cdot a_1) \quad \text{if } r < 0\\
&(a_1 + b_1,\ a_0 + b_0,\ a_0' + b_0',\ a_2 + b_2)
\end{aligned}$$

respectively, for a, b being determined by (a_1, a_0, a_0', a_2), (b_1, b_0, b_0', b_2), respectively.

The other properties of fuzzy numbers and linear fuzzy numbers also have easily derivable analogies for (linear) fuzzy intervals.

For example, the properties of the relations $\sim$, $\succ$, and $\succsim$ are analogous to those in Equations (7.10), (7.11), (7.13), and (7.14). Namely, if a and b are linear fuzzy intervals determined by quadruples (a_1, a_0, a_0', a_2) and (b_1, b_0, b_0', b_2), respectively, then it is easy to verify the following results.

If $[a_0, a_0'] \cap [b_0, b_0'] \neq \emptyset$, then

$$\mu_\sim(a,b) = \mu_\succ(a,b) = \mu_\succ(b,a) = \mu_\succsim(a,b) = \mu_\succsim(b,a) = 1.0$$

If $a_0' < b_0$ and $a_2 > b_1$, then, analogously to the previous section,

$$\begin{aligned}
\mu_\succ(b,a) &= \mu_\succsim(b,a) = 1.0\\
\mu_\succ(a,b) &= \mu_\succsim(a,b) = \mu_\sim(a,b) = \lambda_0
\end{aligned}$$

where $\lambda_0 \in [0,1]$ is such that

$$\lambda_0 = \frac{b_1 - a_2}{a_0' - b_0 + b_1 - a_2}$$

If $a_0' < b_0$ and $a_2 \leq b_1$ then

$$\begin{aligned}
&\mu_\succ(b,a) = \mu_\succsim(b,a) = 1.0\\
&\mu_\sim(a,b) = \mu_\succsim(a,b) = \mu_\succsim(b,a) = \mu_\succ(a,b) = \mu_\succ(b,a) = 0.0
\end{aligned}$$

8

The Convolutive Approach

The definitions of the addition and multiplication operations given in (3.1) and (3.6) are the most common, but not the only ones. Instead of the logical disjunction and conjunction of particular values of the input fuzzy quantities, it is possible to combine them in another way. The convolution of the membership functions seems to be quite natural, and its similarity to the classical probabilistic processing of uncertainty can be rational in those cases in which the modelled uncertainty has some features of both randomness and fuzziness. The convolutive approach to the arithmetics of fuzzy quantities is briefly mentioned, e.g., in [23], and it is widely treated in [34], [59], and [63].

8.1 Convolutive Representation

In this chapter we assume that the membership functions of all fuzzy quantities under consideration are either measurable and integrable, which, together with (2.2) and (2.1), means that the integral

$$\int \mu_a(x)\, dx$$

has meaning for any fuzzy quantity a considered, or the support set (2.3) is countable and the sum

$$\sum \mu_a(x)$$

converges. We will see that specific properties of the operations considered in this chapter can be significantly simplified if in this chapter (and only in this chapter) we replace (2.1) with the following assumption:

$$\mu_a : R \to [0, \infty),\ \mu_a \text{ is limited from above} \tag{8.1}$$

Evidently, (8.1) leads to an even more general concept of fuzzy quantity than does the lattice representation of membership mentioned in the Preface and dealt with in the next chapter. The set of fuzzy quantities fulfilling (8.1) and (2.2) is denoted by $\mathbb{R}^\wedge$.

If $a, b \in \mathbb{R}^\wedge$ then we define their *convolutive sum* $a \boxplus b$ as a fuzzy quantity from $\mathbb{R}^\wedge$ such that for any $x \in R$

$$\mu_{a \boxplus b}(x) = \int \mu_a(y) \cdot \mu_b(x - y)\, dy = \int \mu_a(x - z)\, \mu_b(z)\, dz \qquad (8.2)$$

If the actual model of some numerical data is limited to fuzzy quantities with countable supports, then (8.2) becomes

$$\mu_{a \boxplus b}(x) = \sum \mu_a(y) \cdot \mu_b(x - y) = \sum \mu_a(x - z)\, \mu_b(z) \qquad (8.3)$$

where the summations are realized over the countable union of supports of μ_a and μ_b.

It is easy to see that the operation $\boxplus$ is commutative and associative on $\mathbb{R}^\wedge$, and that $a \boxplus \langle 0 \rangle = a$. The well known difficulties in the case of the usual summation $\oplus$ in connection with the last group property of the opposite element exist in an analogous form even in this case. They can be avoided analogously to the previous case by using the proper equivalence relation instead of the equality. Here, we say that a and b from $\mathbb{R}^\wedge$ are additionally equivalent iff there exist $s_1, s_2 \in \mathbb{S}_0$ such that

$$a \boxplus s_1 = b \boxplus s_2 \qquad (8.4)$$

All important properties of this equivalence, which we denote by $a \sim_{\boxplus} b$, as well as their direct consequences, are closely analogous to the properties of $\sim_{\oplus}$ presented in the previous chapters, as shown, e.g., in [63]. The main source of potential formal difficulties is connected with different levels of membership values of the fuzzy quantities a, b and s_1, s_2. These difficulties can be avoided, in principle, by consequent normalization of the fuzzy quantities in formula (8.4). More details on this subject can be found in [63].

The effectiveness of the described procedure, especially its close analogy with the approach mentioned in Sections 3 and 4, is supported by the fact that convolutive addition preserves symmetry. Let $a, b \in \mathbb{R}^\wedge$, let $u, v \in R$, let a and b be u-symmetric and v-symmetric, respectively. Then, as shown in Section 4, $a = \langle u \rangle \boxplus s_1,\ b = \langle v \rangle \boxplus s_2$ for $s_1, s_2 \in \mathbb{S}_0^\wedge$, and due to the associativity of $\boxplus$

$$a \boxplus b = \langle u + v \rangle \boxplus s_1 \boxplus s_2$$

If $s_1 \boxplus s_2$ is 0-symmetric then $a \boxplus b$ is also $(u + v)$-symmetric. The symmetry of $s_1 \boxplus s_2$ follows from the following relations, valid for any

$x \in R$

$$\mu_{s_1 \boxplus s_2}(x) = \int_R \mu_{s_1}(y)\, \mu_{s_2}(x-y)\, dy = \int_{-\infty}^{\infty} \mu_{s_1}(-y)\, \mu_{s_2}(-x+y)\, dy$$
$$= -\int_{\infty}^{-\infty} \mu_{s_1}(z)\, \mu_{s_2}(-x-z)\, dz = \int_{-\infty}^{\infty} \mu_{s_1}(z)\, \mu_{s_2}(-x-z)\, dz$$
$$= \mu_{s_1 \boxplus s_2}(-x)$$

where the substitution $y = -z$ was used. Analogously, it is possible to show the preservation of symmetry even for the convolutive sum defined by (8.3).

To illustrate the numerical properties of the convolutive addition of fuzzy quantities, it is useful to present a few examples. Since the fuzzy quantities processed below are identical with those dealt with in Examples 4, 5, 6, and 8, it is possible to compare the results of both methods.

Example 26. Let us consider fuzzy quantities $a, b \in \mathbb{R}$ (processed in Example 4) such that

$$\mu_a(1) = \tfrac{1}{3}, \quad \mu_a(2) = \tfrac{2}{3}, \quad \mu_a(3) = 1, \quad \mu_a(x) = 0 \quad \text{for } x \notin \{1,2,3\}$$
$$\mu_b(4) = \tfrac{1}{2}, \quad \mu_b(5) = 1, \quad \mu_b(6) = \tfrac{1}{2}, \quad \mu_b(x) = 0 \quad \text{for } x \notin \{4,5,6\}$$

Then, using (8.3), we can derive

$$\mu_{a \boxplus b}(5) = \mu_a(1) \cdot \mu_b(4) = \frac{1}{3} \cdot \frac{1}{2} = \frac{1}{6}$$
$$\mu_{a \boxplus b}(6) = \mu_a(1) \cdot \mu_b(5) + \mu_a(2) \cdot \mu_b(4) = \frac{2}{3}$$

and further, analogously,

$$\mu_{a \boxplus b}(7) = \frac{1}{3} \cdot \frac{1}{2} + \frac{2}{3} \cdot 1 + 1 \cdot \frac{1}{2} = \frac{4}{3}$$
$$\mu_{a \boxplus b}(8) = \frac{4}{3}, \quad \mu_{a \boxplus b}(9) = \frac{1}{2}, \quad \mu_{a \boxplus b}(x) = 0 \quad \text{for other } x \in R$$

Formulas (8.2) and (8.3) do not admit the addition of continuous and discrete fuzzy quantities. Nevertheless, the heuristic background and inner logic of those relations fully justify the procedure described by

$$\mu_{a \boxplus b}(x) = \sum_{y \in M} \mu_a(x-y) \cdot \mu_b(y), \quad x \in R$$

where b is the discrete component of the sum and M is the discrete support of μ_b,

$$M = \{x \in R : \mu_b(x) > 0\}$$

Meanwhile, the support of a is an uncountable set (e.g., union of intervals). Then the convolutive sum of the fuzzy quantities dealt with in Example 5 can be calculated as shown in the following example.

Example 27. If a, $b \in \mathbb{R}$ are fuzzy quantities,

$$\begin{aligned} \mu_a(y) &= y-1 \quad \text{for } y \in [1,2], \\ &= 3-y \quad \text{for } y \in [2,3], \\ &= 0 \qquad\ \text{for } y \notin [1,3], \end{aligned}$$

$$\mu_b(4) = 1, \quad \mu_b(5) = \tfrac{1}{2}, \quad \mu_b(z) = 0 \quad \text{for } z \notin \{4,5\}$$

Then

$$\mu_{a \boxplus b}(x) = 0 \quad \text{for } x \notin [5,8]$$

since for such x and for any $z \in R$ at least one of the values $\mu_a(x - z)$, $\mu_b(z)$ is equal to 0. Further, for $x \in [5,6]$

$$\mu_{a \boxplus b}(x) = \mu_a(x-4) \cdot \mu_b(4) = \mu_a(x-4) = x-5$$

For $x \in [6,7]$

$$\begin{aligned} \mu_{a \boxplus b}(x) &= \mu_a(x-4) \cdot \mu_b(4) + \mu_a(x-5) \cdot \mu_b(5) \\ &= (7-x) \cdot 1 + \frac{1}{2} \cdot (x-6) = 4 - \frac{1}{2}x \end{aligned}$$

and for $x \in [7,8]$

$$\mu_{a \boxplus b}(x) = \mu_a(x-5) \cdot \mu_b(5) = (8-x) \cdot \frac{1}{2} = 4 - \frac{1}{2}x$$

The addition of continuous fuzzy quantities is defined by the integral formula (8.2). In the simpler case mentioned in Example 6, the computation proceeds as follows.

Example 28. Let us consider fuzzy quantities a, $b \in \mathbb{R}$ with

$$\begin{aligned} \mu_a(y) &= y-1 \quad \text{for } y \in [1,2], \\ &= 0 \qquad\ \text{for } y \notin [1,2], \\ \mu_b(z) &= 4-z \quad \text{for } z \in [3,4], \\ &= 0 \qquad\ \text{for } z \notin [3,4] \end{aligned}$$

Then

$$\int_R \mu_a(y)\,\mu_b(x-y)\,dy = \int_1^2 (y-1)\cdot\mu_b(x-y)\,dy = \int_M (y-1)\cdot(4-x+y)\,dy$$

where $M = [1,2] \cap [x-4,\ x-3]$.

For $x < 4$ or $x > 6$ the set M is empty, and the integral is equal to 0.

If $x \in [4,5]$ then $M = [1,\ x-3]$ and

$$\int_1^{x-3} (y-1)\,(4-x+y)\,dy = -\frac{1}{3}\,x^3 + \frac{5}{2}\,x^2 - 12x + \frac{56}{3}$$

hence, $\mu_{a\boxplus b}(4) = 0,\ \mu_{a\boxplus b}(5) = \frac{2}{3}$.

If $x \in [5,6]$ then $M = [x-4,\ 2]$ and

$$\int_{x-4}^{2} (y-1)\,(4-x+y)\,dy = \frac{1}{6}\,x^3 - \frac{5}{2}\,x^2 + 12x - 18$$

hence $\mu_{a\boxplus b}(5) = \frac{2}{3},\ \mu_{a\boxplus b}(6) = 0$.

We can see that the membership function $\mu_{a\boxplus b}$ is continuous.

More complicated computation is connected with the next, rather more complex, fuzzy quantities.

Example 29. Let $a,\ b \in \mathbb{R}$ be fuzzy quantities with membership functions

$$\begin{aligned} \mu_a(y) &= y-1 && \text{for } y \in [1,2), & \mu_b(z) &= 5-z && \text{for } z \in [4,5],\\ &= 1 && \text{for } y \in [2,3), & &= z-5 && \text{for } z \in [5,6],\\ &= 0 && \text{for } y \notin [1,3), & &= 0 && \text{for } z \notin [4,6] \end{aligned}$$

(see Example 8). Then, using (8.2), we can calculate

$$\begin{aligned} \mu_{a\boxplus b}(x) &= \int_R \mu_a(y)\,\mu_b(x-y)\,dy\\ &= \int_{M_1} (y-1)\cdot(5-x+y)\,dy + \int_{M_2} (y-1)\,(x-y-5)\,dy\\ &\quad + \int_{M_3} (5-x+y)\,dy + \int_{M_4} (x-y-5)\,dy \end{aligned}$$

where

$$M_1 = [1,2] \cap [x-5,\ x-4], \quad M_2 = [1,2] \cap [x-6,\ x-5]$$

$$M_3 = [2,3) \cap [x-5,\ x-4], \quad M_4 = [2,3) \cap [x-6,\ x-5]$$

If $x < 5$ or $x > 9$, then $M_1,\ M_2,\ M_3,$ *and* M_4 are empty or one-element sets and, consequently, $\mu_{a\boxplus b}(x) = 0$.

Let $x \in [5,6]$, then $M_1 = [1, x-4]$, $M_2 \subset \{1\}$, $M_3 \subset \{2\}$, $M_4 = \emptyset$ and

$$\mu_{a \boxplus b}(x) = \int_1^{x-4} (y-1)\,(5-x+y)\,dy = \frac{1}{6}\,x^3 + 3x^2 - \frac{35}{2}\,x + \frac{100}{3}$$

which also means that $\mu_{a \boxplus b}(5) = 0,\ \mu_{a \boxplus b}(6) = \frac{1}{3}$.

If $x \in [6,7]$ then $M_1 = [x-5,\, 2]$, $M_2 = [1,\, x-5]$, $M_3 = [2,\, x-4]$, $M_4 = \emptyset$ and

$$\mu_{a \boxplus b}(x) = \int_{x-5}^{2} (y-1)\,(5-x+y)\,dy + \int_1^{x-5} (y-1)\,(x-y-5)\,dy$$
$$+ \int_2^{x-4} (5-x+y)\,dy = \frac{1}{3}\,x^3 - \frac{12}{2}\,x^2 + \frac{85}{2}\,x - \frac{278}{3}$$

This also means that $\mu_{a \boxplus b}(6) = \frac{1}{3},\ \mu_{a \boxplus b}(7) = \frac{2}{3}$.

If $x \in [7,8]$, then $M_1 \subset \{2\}$, $M_2 = [x-6,\, 2]$, $M_3 = [x-5, 3)$, $M_4 = [2,\, x-5]$, and

$$\mu_{a \boxplus b}(x) = \int_{x-6}^{2} (y-1)\,(x-y-5)\,dy + \int_{x-5}^{3} (5-x+y)\,dy$$
$$+ \int_2^{x-5} (x-y-5)\,dy = -\frac{x^3}{6} + 4x^2 - 32x + \frac{515}{6}$$

The marginal values are $\mu_{a \boxplus b}(7) = \frac{2}{3},\ \mu_{a \boxplus b}(8) = \frac{1}{2}$.

Finally, if $x \in [8,9]$, then $M_1 = \emptyset$, $M_2 \subset \{2\}$, $M_3 = \emptyset$, $M_4 = [x-6,\, 3)$, and

$$\mu_{a \boxplus b}(x) = \int_{x-6}^{3} (x-y+5)\,dy = \frac{x^2}{2} + 8x - \frac{63}{2}$$

with $\mu_{a \boxplus b}(8) = \frac{1}{2},\ \mu_{a \boxplus b}(9) = 0$.

The polynomial forms of $\mu_{a \boxplus b}$ for particular intervals and the coincidence of the values of $\mu_{a \boxplus b}$ for the marginal points show that the membership function $\mu_{a \boxplus b}$ is continuous on R.

Similarly, the *convolutive product* of $a,\, b \in \mathbb{R}^\wedge$, denoted by $a \boxdot b \in \mathbb{R}^\wedge$, is defined by

$$\mu_{a \boxdot b}(x) = \int \mu_a(y)\,\mu_b(x/y)\,dy = \int \mu_a(x/z)\,\mu_b(z)\,dz \qquad (8.5)$$

where (2.2) allows integration over the set of non-zero real numbers. If this model is applied to fuzzy quantities with countable supports, then

$$\mu_{a \boxdot b}(x) = \sum_{y \neq 0} \mu_a(y)\,\mu_b(x/y) = \sum_{z \neq 0} \mu_a(x/z)\,\mu_b(z) \qquad (8.6)$$

Even in this case the properties of this convolutive fuzzy product are closely analogous to those of the former operation, $a \odot b$.

We can illustrate the numerical calculation based on (8.5) and (8.6) with a few examples. As we use some of the fuzzy quantities dealt with in some of the examples in Section 3, the comparison of results can offer an interesting view of the properties of the operations $\odot$ and $\boxdot$, and of the comparison of their results. Let us start with the product of two discrete fuzzy quantities.

Example 30. Let $a,\, b \in \mathbb{R}$ be fuzzy quantities with

$$\mu_a(1) = \tfrac{1}{3}, \quad \mu_a(2) = \tfrac{2}{3}, \quad \mu_a(3) = 1, \quad \mu_a(y) = 0 \quad \text{for } y \notin \{1,2,3\},$$
$$\mu_b(4) = \tfrac{1}{2}, \quad \mu_b(5) = 1, \quad \mu_b(6) = \tfrac{1}{2}, \quad \mu_b(z) = 0 \quad \text{for } z \notin \{4,5,6\}$$

Then, using (8.6), we obtain

$$\mu_{a\boxdot b}(4) = \mu_a(1) \cdot \mu_b(4) = \frac{1}{3} \cdot \frac{1}{2} = \frac{1}{6},$$
$$\mu_{a\boxdot b}(5) = \frac{1}{3} \cdot 1 = \frac{1}{3}, \quad \mu_{a\boxdot b}(6) = \frac{1}{3} \cdot \frac{1}{2} = \frac{1}{6}, \quad \mu_{a\boxdot b}(8) = \frac{2}{3} \cdot \frac{1}{2} = \frac{1}{3},$$
$$\mu_{a\boxdot b}(10) = \frac{2}{3} \cdot 1 = \frac{2}{3}, \quad \mu_{a\boxdot b}(12) = \frac{2}{3} \cdot \frac{1}{2} = \frac{1}{3}, \quad \mu_{a\boxdot b}(15) = 1 \cdot 1 = 1,$$
$$\mu_{a\boxdot b}(18) = 1 \cdot \frac{1}{2} = \frac{1}{2}, \quad \mu_{a\boxdot b}(x) = 0 \quad \text{for other } x \in R$$

In the case of product of continuous and discrete fuzzy quantities not formally covered by formulas (8.5) and (8.6), we can proceed analogously to the additive case presented in Example 27. It is natural to define for $a,\, b \in \mathbb{R}_0$, where the support set of b is discrete, their convolutive product $a \boxdot b$ by

$$\mu_{a\boxdot b}(x) = \sum_{y \in M} \mu_a(x/y)\, \mu_b(y), \qquad x \in R$$

where $M \subset R_0$ is the discrete support set of μ_b. In this way we proceed with the following example.

Example 31. Let $a,\, b \in \mathbb{R}$ be such that

$$\begin{aligned} \mu_a(y) &= y - 1 && \text{for } y \in [1,2], \\ &= 3 - y && \text{for } y \in [2,3], \\ &= 0 && \text{for } y \notin [1,3], \end{aligned}$$
$$\mu_b(4) = 1, \quad \mu_b(5) = \tfrac{1}{2}, \quad \mu_b(z) = 0 \quad \text{for } z \notin \{4,5\}$$

If we accept the modification of (8.5) and (8.6)

$$\mu_{a\Box b}(x) = \mu_a\left(\frac{x}{4}\right) + \frac{1}{2}\,\mu_a\left(\frac{x}{5}\right)$$

then

$$\mu_{a\Box b}(x) = 0 \quad \text{for } x \notin [4, 15]$$

since for such x both $\mu_a\left(\frac{x}{4}\right)$ and $\mu_a\left(\frac{x}{5}\right)$ vanish.

If $x \in [4, 5]$, then

$$\mu_a\left(\frac{x}{5}\right) = 0 \quad \text{and} \quad \mu_{a\Box b}(x) = \frac{x}{4} - 1$$

and also $\mu_{a\Box b}(4) = 0$, $\mu_{a\Box b}(5) = \frac{1}{4}$.

If $x \in [5, 8]$, then

$$\mu_{a\Box b}(x) = \frac{x}{4} - 1 + \frac{1}{2}\left(\frac{x}{5}1\right) = \frac{7}{20}\,x - \frac{3}{2}$$

and $\mu_{a\Box b}(5) = \frac{1}{4}$, $\mu_{a\Box b}(8) = \frac{13}{10}$.

If $x \in [8, 10]$, then

$$\mu_{a\Box b}(x) = 3 - \frac{x}{4} + \frac{1}{2}\left(\frac{x}{5} - 1\right) = \frac{5}{2} - \frac{3}{20}\,x$$

and $\mu_{a\Box b}(8) = \frac{13}{10}$, $\mu_{a\Box b}(10) = 1$.

If $x \in [10, 12]$, then

$$\mu_{a\Box b}(x) = 3 - \frac{x}{4} + \frac{1}{2}\left(3 - \frac{x}{5}\right) = \frac{9}{2} - \frac{7}{20}\,x$$

and $\mu_{a\Box b}(10) = 1$, $\mu_{a\Box b}(12) = \frac{3}{10}$.

If $x \in [12, 15]$, then $\mu_a\left(\frac{x}{4}\right) = 0$,

$$\mu_{a\Box b}(x) = \frac{1}{2}\left(3 - \frac{x}{5}\right) = \frac{3}{2} - \frac{x}{10}$$

and $\mu_{a\Box b}(12) = \frac{3}{10}$, $\mu_{a\Box b}(15) = 0$.

This means that $\mu_{a\Box b}$ is a continuous function.

The convolutive product of fuzzy quantities with continuous membership functions presented in Example 12 also illustrates the properties of the operation $\boxdot$.

Example 32. Let a, $b \in \mathbb{R}$ be fuzzy quantities with

$$\begin{aligned} \mu_a(y) &= y - 1 \quad \text{for } y \in [1, 2], & \mu_b(z) &= 4 - z \quad \text{for } z \in [3, 4], \\ &= 0 \quad \text{for } y \notin [1, 2], & &= 0 \quad \text{for } z \notin [3, 4]. \end{aligned}$$

Then, analogously to other similar examples,

$$\mu_{a\Box b}(x) = \int_M (y-1)\left(4 - \frac{x}{y}\right) dy$$

where $M = [1, 2] \cap \left[\frac{x}{4}, \frac{x}{3}\right]$ as follows from (8.5) Obviously $M = \emptyset$ for $x < 3$ and for $x > 8$. Hence,

$$\mu_{a\Box b}(x) = 0 \quad \text{for } x \notin [3, 8]$$

If $x \in [3, 4]$, then $M = \left[1, \frac{x}{3}\right]$ and

$$\mu_{a\Box b}(x) = \int_1^{\frac{x}{3}} (y-1)\left(4 - \frac{x}{y}\right) dy = \frac{1}{9}\,x^2 + x\cdot\left(\ln\frac{x}{3} - \frac{1}{3}\right) + 2$$

i.e., $\mu_{a\Box b}(3) = 0,\ \mu_{a\Box b}(4) \doteq 0.0396$.

If $x \in [4, 6]$ then $M = \left[\frac{x}{4}, \frac{x}{3}\right]$ and

$$\mu_{a\Box b}(x) = \frac{1}{72}\,x^2 + x\cdot\left(\ln\frac{4}{3}\frac{1}{3}\right)$$

i.e., $\mu_{a\Box b}(4) \doteq 0.0396,\ \mu_{a\Box b}(6) \doteq 0.226$.

For $x \in [6, 8]$ the set M is equal to $\left[\frac{x}{4}, 2\right]$, and

$$\mu_{a\Box b}(x) = \frac{x^2}{8} + x\cdot\left(\ln\frac{8}{x}1\right)$$

i.e., $\mu_{a\Box b}(6) \doteq 0.226,\ \mu_{a\Box b}(8) = 0$.

Some interesting features of the convolutive products can also be seen in the remaining two examples in this section.

The convolutive product defined above lacks, in general, one very useful property. Namely, formula (8.5) does not preserve the transversibility of the input fuzzy quantities. Let $a, b \in \mathbb{R}^\wedge$ be 1-transversible. Then for any $x \in R_0$

$$\begin{aligned}
\mu_{a\Box b}\left(\frac{1}{x}\right) &= \int_{-\infty}^{\infty} \mu_a(y)\,\mu_b\left(\frac{1}{xy}\right) dy \\
&= \int_{-\infty}^{\infty} \mu_a\left(\frac{1}{y}\right)\,\mu_b(xy)\,dy = \int_{-\infty}^{\infty} \frac{1}{x}\,\mu_a\left(\frac{x}{z}\right)\,\mu_b(z)\,dz \\
&= \frac{1}{x}\,\mu_{a\Box b}(x)
\end{aligned}$$

where the substitution $x/y = z$ was used. The following example illustrates this fact. If the input fuzzy quantities are discrete and (8.6) is used, then the discrepancy mentioned above does not appear. This can be derived from (8.6) analogously to the previous procedure, and it is illustrated by Example 34.

Example 33. Let us consider a 1-transversible fuzzy quantity $a \in R$ with

$$\begin{aligned} \mu_a(y) &= 2y - 1 \quad \text{for } y \in \left[\tfrac{1}{2}, 1\right] \\ &= \tfrac{2}{y} - 1 \quad \text{for } y \in [1, 2] \\ &= 0 \quad \text{for } y \notin \left[\tfrac{1}{2}, 2\right] \end{aligned}$$

Then we can calculate the product $a \odot a$ using

$$\mu_{a \odot a}(x) = \sup_{y \in R_0} \left(\min \left(\mu_a(y), \, \mu_a\left(\frac{x}{y}\right)\right)\right)$$

If $x \notin \left[\frac{1}{4}, 4\right]$ then for any $y \in R_0$ at least one of the values $\mu_a(y)$ and $\mu_a\left(\frac{x}{y}\right)$ is equal to 0 and, consequently, $\mu_{a \odot a}(x) = 0$.

If $x \in \left[\frac{1}{4}, 1\right]$ then it can be easily verified that

$$\sup_{y \in R_0} \left(\min \left(\mu_a(y), \, \mu_a\left(\frac{x}{y}\right)\right)\right) = \max_{y \in \left[\frac{1}{2}, 1\right]} \left(\min \left(2y - 1, \, 2\frac{x}{y} - 1\right)\right)$$

This maximum is achieved for the value $y \in \left[\frac{1}{2}, 1\right]$, for which

$$\mu_a(y) = \mu_a\left(\frac{x}{y}\right)$$

i.e.,

$$2y - 1 = 2\,\frac{x}{y} - 1 \quad \text{and consequently} \quad y = \sqrt{x}$$

For this y

$$\mu_a(y) = \mu_a\left(\frac{x}{y}\right) = 2\sqrt{x} - 1$$

On the other hand, if $x \in [1, 4]$, then analogously

$$\sup_{y \in R_0} \left(\min \left(\mu_a(y), \, \mu_a\left(\frac{x}{y}\right)\right)\right) = \max_{y \in [1, 2]} \left(\min \left(\frac{2}{y} - 1, \, \frac{2y}{x} - 1\right)\right)$$

This maximum is achieved for the $y \in [1, 2]$ for which $\mu_a(y) = \mu_a\left(\frac{x}{y}\right)$ — it is the y for which

$$\frac{2}{y} - 1 = \frac{2y}{x} - 1, \quad \text{i.e.,} \quad y = \sqrt{x}$$

For this y

$$\mu_a(y) = \mu_a\left(\frac{x}{y}\right) = \frac{2}{\sqrt{x}} - 1 = \frac{2\sqrt{x}}{x} - 1$$

This means that $\mu_{a\odot a}$ is a continuous function

$$\begin{aligned}\mu_{a\odot a}(x) &= 2\sqrt{x}-1 && \text{for } x\in\left[\tfrac{1}{4},\,1\right]\\ &= \tfrac{2}{\sqrt{x}}-1 && \text{for } x\in[1,4]\\ &= 0 && \text{for } x\notin\left[\tfrac{1}{4},\,4\right]\end{aligned}$$

and is evidently 1-transversible.

The computation of the convolutive product $a \boxdot a$ is also relatively easy. For any $x \in R$

$$\int_{R_0}\mu_a(y)\cdot\mu_a\left(\frac{x}{y}\right)\,dy+\int_{M_1}(2y-1)\cdot\left(2\frac{x}{y}-1\right)\,dy$$
$$+\int_{M_2}(2y-1)\cdot\left(\frac{2y}{x}-1\right)\,dy+\int_{M_3}\left(\frac{2}{y}-1\right)\cdot\left(2\frac{x}{y}-1\right)\,dy$$
$$+\int_{M_4}\left(\frac{2}{y}-1\right)\cdot\left(\frac{2y}{x}-1\right)\,dy$$

where

$$M_1=\left[\tfrac{1}{2},1\right]\cap[x,\,2x],\quad M_2=\left[\tfrac{1}{2},1\right]\cap\left[\tfrac{1}{2}x,\,x\right]$$
$$M_3=[1,2]\cap[x,2x],\qquad M_4=[1,2]\cap\left[\tfrac{1}{2}x,\,x\right]$$

Then, analogously to Example 32, e.g., $\mu_{a\boxdot a}(x)=0$ for $x\notin\left[\frac{1}{4},\,4\right]$. If $x\in\left[\frac{1}{4},\,\frac{1}{2}\right]$ then $M_1=\left[\frac{1}{2},\,2x\right]$, $M_2\subset\left\{\frac{1}{2}\right\}$, $M_3\subset\{1\}$, $M_4=\emptyset$, and

$$\mu_{a\boxdot a}(x)=\int_{\frac{1}{2}}^{2x}(2y-1)\cdot\left(2\frac{x}{y}-1\right)\,dy=4x^2-2x\ln 4x-\frac{1}{4}$$

where $\mu_{a\boxdot a}\left(\frac{1}{4}\right)=0$, $\mu_{a\boxdot a}\left(\frac{1}{2}\right)=\frac{3}{4}-\ln 2\doteq 0.0568$.

If $x\in\left[\frac{1}{2},\,1\right]$, then $M_1=[x,1]$, $M_2=\left[\frac{1}{2},\,x\right]$, $M_3=[1,2x]$, $M_4\subset\{1\}$. This means that

$$\begin{aligned}\mu_{a\boxdot a}(x)&=\int_x^1(2y-1)\cdot\left(2\frac{x}{2}-1\right)\,dy+\int_{\frac{1}{2}}^{x}(2y-1)\cdot\left(\frac{2y}{x}-1\right)\,dy\\&\quad+\int_1^{2x}\left(\frac{2}{y}-1\right)\cdot\left(2\frac{x}{y}-1\right)\,dy\\&=-\frac{8}{3}x^2+9x-\frac{13}{4}+\frac{1}{12x}+2x\ln x-(2+2x)\ln 2x\end{aligned}$$

and $\mu_{a\boxdot a}\left(\frac{1}{2}\right)=\frac{3}{4}+\ln\frac{1}{2}\doteq 0.0568$, $\mu_{a\boxdot a}(1)=\frac{19}{6}-4\cdot\ln 2\doteq 0.394$.

If $x\in[1,2]$ then $M_1\subset\{1\}$, $M_2=\left[\frac{x}{2},1\right]$, $M_3=[x,2]$, $M_4=[1,x]$ and

$$\mu_{a\boxdot a}(x)=\int_{\frac{x}{2}}^{1}(2y-1)\cdot\left(\frac{2y}{x}-1\right)\,dy+\int_x^2\left(\frac{2}{y}-1\right)\cdot\left(2\frac{x}{y}-1\right)\,dy$$

$$+\int_1^x \left(\frac{2}{y}-1\right)\cdot\left(\frac{2y}{x}-1\right)\,dy$$

$$=\frac{1}{12}x^2-\frac{13}{4}x+9-\frac{8}{3x}+2\ln\frac{1}{x}-(2+2x)\ln\frac{2}{x}$$

Here $\mu_{a\Box a}(1)=\frac{19}{6}-4\ln 2\doteq 0.394$, $\mu_{a\Box a}(2)=\frac{3}{2}-2\ln 2\doteq 0.1137$.

If, finally, $x\in[2,4]$, then $M_1=\emptyset$, $M_2\subset\{1\}$, $M_3\subset\{2\}$, $M_4=\left[\frac{x}{2},2\right]$. This means that

$$\mu_{a\Box a}(x)=\int_{\frac{x}{2}}^2\left(\frac{2}{y}-1\right)\cdot\left(\frac{2y}{x}-1\right)\,dy=\frac{x}{4}+\frac{4}{x}-2\ln\frac{4}{x}$$

and $\mu_{a\Box a}(2)=\frac{3}{2}-2\ln 2=0.1137$, $\mu_{a\Box a}(4)=0$.

These results illustrate the previous general conclusion regarding the relation between $\mu_{a\Box a}(x)$ and $\mu_{a\Box a}\left(\frac{1}{x}\right)$. Even if a is 1-transversible and the classical product $a\odot a$ preserves the 1-transversibility, the convolutive product $\mu_{a\Box a}$ is not transversible at all.

We can see that the convolutive product of discrete fuzzy quantities using (8.6) preserves transversibility, as illustrated by the following example.

Example 34. Let a, $b\in\mathbb{R}$ be fuzzy quantities such that

$$\begin{aligned}
&\mu_a\left(\tfrac{1}{4}\right)=\tfrac{1}{2}, \quad \mu_a\left(\tfrac{1}{2}\right)=1, \quad \mu_a(1)=\tfrac{1}{2}, \quad \mu_a(x)=0 \quad \text{for other } x\in R,\\
&\mu_b\left(\tfrac{2}{3}\right)=\tfrac{1}{3}, \quad \mu_b(1)=\tfrac{2}{3}, \quad \mu_b\left(\tfrac{4}{3}\right)=1, \quad \mu_b(2)=\tfrac{2}{3},\\
&\mu_b(3)=1, \quad \mu_b(4)=\tfrac{2}{3}, \quad \mu_b(6)=\tfrac{1}{3}, \quad \mu_b(x)=0 \quad \text{for other } x\in R.
\end{aligned}$$

It is easy to verify that a is $\frac{1}{2}$-transversible and that b is 2-transversible. Their convolutive product calculated by means of (8.6) is

$$\begin{aligned}
\mu_{a\Box b}\left(\tfrac{1}{6}\right)&=\mu_a\left(\tfrac{1}{4}\right)\cdot\mu_b\left(\tfrac{2}{3}\right)=\tfrac{1}{2}\cdot\tfrac{1}{3}=\tfrac{1}{6},\\
\mu_{a\Box b}\left(\tfrac{1}{4}\right)&=\mu_a\left(\tfrac{1}{4}\right)\cdot\mu_b(1)=\tfrac{1}{2}\cdot\tfrac{2}{3}=\tfrac{1}{3},\\
\mu_{a\Box b}\left(\tfrac{1}{3}\right)&=\mu_a\left(\tfrac{1}{4}\right)\cdot\mu_b\left(\tfrac{4}{3}\right)+\mu_a\left(\tfrac{1}{2}\right)\cdot\mu_b\left(\tfrac{2}{3}\right)=\tfrac{1}{2}\cdot 1+1\cdot\tfrac{1}{3}=\tfrac{5}{6},
\end{aligned}$$

and analogously

$$\mu_{a\Box b}\left(\frac{1}{2}\right)=1, \quad \mu_{a\Box b}\left(\frac{2}{3}\right)=\frac{7}{6}, \quad \mu_{a\Box b}\left(\frac{3}{4}\right)=\frac{1}{2},$$

$$\mu_{a\Box b}(1)=\mu_a\left(\frac{1}{4}\right)\cdot\mu_b(4)+\mu_a\left(\frac{1}{2}\right)\cdot\mu_b(2)=\mu_a(1)\cdot\mu_b(1)=\frac{4}{3},$$

$$\mu_{a\Box b}\left(\frac{4}{3}\right)=\frac{1}{2}, \quad \mu_{a\Box b}\left(\frac{3}{2}\right)=\frac{7}{6}, \quad \mu_{a\Box b}(2)=1,$$

$$\mu_{a\boxdot b}(3) = \frac{5}{6}, \quad \mu_{a\boxdot b}(4) = \frac{1}{3}, \quad \mu_{a\boxdot b}(6) = \frac{1}{6},$$

$$\mu_{a\boxdot b}(x) = 0 \quad \text{for other } x \in R.$$

Obviously, the fuzzy quantity $a \boxdot b$ is 1-transversible.

Of course, the concept of the crisp product $r \cdot a$ of $r \in R$, $a \in \mathbb{R}^\wedge$ and its properties remain unchanged.

In general, the convolutive approach can be applied to all operations covered by the representation theorem. Its classical form mentioned in Section 2.3, in (2.13), can be modified, using the convolution operation, in the following way. If $f : R \times R \to R$ is a binary operation, then we extend it to the operation over $\mathbb{R}^\wedge$, and for $a, b, c \in \mathbb{R}^\wedge$, $c = f(a, b)$ is defined by

$$\mu_c(z) = \int_M \mu_a(x) \cdot \mu_b(y)\, dx dy, \quad z \in R,\ M = \{(x, y) \in R^2 : z = f(x, y)\} \tag{8.7}$$

provided that the integral has meaning. If we consider fuzzy quantities with a countable support Q (e.g., rational or integral numbers), then (2.13) turns into the convolutive formula

$$\mu_c(z) = \sum_M \mu_a(x)\, \mu_b(y), \quad \text{where } z \in Q,\ M = \{(x, y) \in Q^2,\ z = f(x, y)\} \tag{8.8}$$

defining the membership function μ_c of $c = f(a, b)$.

8.2 Specificity of Convolutive Approach

If we compare the general formulas (8.7) and (2.13), or their special cases (8.2) and (3.1) or (8.5) and (3.7), their formal similarity is easily seen. It is natural: in both approaches, the convolutive and the supremum-minimum one, we combine the values of membership functions belonging to the arguments, which together form the value of the result of some arithmetic operation. This common structure of the two approaches implies that both may be and sometimes are used, in correspondence with the inner structure of the modeled fuzzy situation. Even if the supremum-minimum representation appears more frequently it can be useful to compare some of their advantages and disadvantages.

At any rate, the similarity of formalisms does not fully exclude some differences that can be rather surprising. The main discrepancy of the convolutive approach was illustrated in the previous section. It consists in the fact that the convolutive product does not preserve transversibility, as shown also in Example 30 and in the theoretical paragraph illustrated by that example. But some properties of the convolutive approach can even be advantageous.

The convolutive representation of arithmetic operations can sometimes be more lucid if some theoretical properties are to be derived. The theory of integrals simply offers more formal tools than does the processing and combining of extremes. On the other hand — if the convolutive representation is used then it is desirable to introduce and verify several formal assumptions concerning the measurability and integrability of the membership functions.

As the integrals (8.2) and (8.5) or the sums (8.3) and (8.6) can generally be greater than 1, the membership values cannot easily be limited by the interval $[0, 1]$. In such a case we can decide between two general strategies. Either we accept the membership functions with values in the interval $[0, \infty)$, as was done in (8.1). In such case we interpret the memberships $\mu_a(x) > 1$ as the certainty that the quantity a can achieve the value x (this is similar to $\mu_a(x) = 1$). Or, we can somehow normalize the values $\mu_{a \boxplus b}(x) > 1$ or $\mu_{a \boxdot b}(x) > 1$ after each realization of the relevant operation $a \boxplus b$ or $a \boxdot b$. We can truncate the superfluous membership values by 1 or divide the values $\mu_a(x)$, $x \in \mathbb{R}$, by their supremal value

$$\sup(\mu_a(y) : y \in R) \tag{8.9}$$

Both methods, truncation and normalization, can cause serious difficulties by breaking the associativity of the operations $\boxplus$ and $\boxdot$. It means that in order to use them in some concrete applied algorithm it is necessary to declare exactly the sequence of its particular steps so that different results from the same problem are not obtained by performing the arithmetic operations in a different order. In the case of concrete and formally closed algorithms, it is possible to truncate or normalize only the final result of computation when all previous current semi-results were left untruncated and unnormalized. Regarding the interpretation of such results, it is similar to accepting the values $\mu_a(x) > 1$ and interpreting them as $\mu_a(x) = 1$, as mentioned above.

The above discrepancy of the convolutive approach is compensated for by another of its analytical properties. It is obvious that each application of the operations $\boxplus$ and $\boxdot$ also means a combination of the support sets of the input fuzzy quantities. Generally, the support set of $a \boxplus b$ is in some sense richer — and wider — than the support sets of a and b. If we denote by x_1, y_1, z_1 the numbers

$$x_1 = \sup(x_0 \in R : \mu_a(x) = 0 \text{ for all } x < x_0) \tag{8.10}$$

and analogously y_1 for b and z_1 for $a \boxplus b$, and if we similarly denote by x_2, y_2, z_2 the values

$$x_2 = \inf(x_0 \in R : \mu_a(x) = 0 \text{ for all } x > x_0) \tag{8.11}$$

with analogous definition of y_2 for b and z_2 for $a \boxplus b$, then

$$(z_2 - z_1) \geq \max\left[(x_2 - x_1), (y_2 - y_1)\right]$$

Moreover,

$$z_2 - z_1 = (x_2 - x_1) + (y_2 - y_1)$$

A similar, even if not exactly identical, conclusion is valid for the convolutive multiplication.

The same relations can be derived for the supremum-minimum sum $a \oplus b$ and product $a \odot b$. This means that repetitive application of both approaches enormously enlarges the range of possible values of the output fuzzy quantities. But there is a difference that can be significant, especially in the case of discrete (or even finite) supports. Meanwhile in the classical supremum-minimum procedures, the possibility (i.e., the membership function value) of the marginal elements of this range cannot be less than the minimal possibilities of the input quantities — otherwise, the repetitive application of the convolutive procedure leads to the vanishing of the possibility of the marginal values. This offers a possibility of compensating for the enlarging of the support set by omitting its marginal, less possible elements. This can be illustrated by the following example, and even by the comparison of some examples presented in the previous sections. Specifically, it is useful to compare the results of the following pairs of examples dealing with the same fuzzy quantities:

Examples 4 and 26
Examples 5 and 27
Examples 6 and 28
Examples 8 and 29

in the additive case, and

Examples 10 and 30
Examples 11 and 31
Examples 12 and 32

in the multiplicative case. The following simple example illustrates the same conclusion as well.

Example 35. Let us consider a fuzzy quantity $a \in \mathbb{R} \subset \mathbb{R}^\wedge$ such that

$$\mu_a(1) = 0.5, \quad \mu_a(2) = 1.0, \quad \mu_a(3) = 0.2, \quad \mu_a(x) = 0 \quad \text{for other } x.$$

Then it is easy to compute, using (3.1),

$$\mu_{a \oplus a}(2) = 0.5, \quad \mu_{a \oplus a}(3) = 0.5, \quad \mu_{a \oplus a}(4) = 1.0,$$
$$\mu_{a \oplus a}(5) = 0.2, \quad \mu_{a \oplus a}(6) = 0.2, \quad \mu_{a \oplus a}(x) = 0 \quad \text{for other } x,$$

and

$$\mu_{a\oplus a\oplus a}(3) = \mu_{a\oplus a\oplus a}(4) = \mu_{a\oplus a\oplus a}(5) = 0.5,$$

$$\mu_{a\oplus a\oplus a}(6) = 1.0, \qquad \mu_{a\oplus a\oplus a}(7) = \mu_{a\oplus a\oplus a}(8) = \mu_{a\oplus a\oplus a}(9) = 0.2,$$

$$\mu_{a\oplus a\oplus a}(x) = 0 \quad \text{for other } x \in R.$$

But the convolutive formula (8.3) results in the following values

$$\mu_{a\boxplus a}(2) = 0.25, \quad \mu_{a\boxplus a}(3) = 1.0, \quad \mu_{a\boxplus a}(4) = 1.2,$$

$$\mu_{a\boxplus a}(5) = 0.4, \quad \mu_{a\boxplus a}(6) = 0.04, \quad \mu_{a\boxplus a}(x) = 0 \quad \text{for other } x.$$

The development of the membership values continues for $a \boxplus a \boxplus a$, as follows:

$$\mu_{a\boxplus a\boxplus a}(3) = 0.125, \quad \mu_{a\boxplus a\boxplus a}(4) = 0.75, \quad \mu_{a\boxplus a\boxplus a}(5) = 1.65,$$

$$\mu_{a\boxplus a\boxplus a}(6) = 1.6, \quad \mu_{a\boxplus a\boxplus a}(7) = 0.84, \quad \mu_{a\boxplus a\boxplus a}(8) = 0.12,$$

$$\mu_{a\boxplus a\boxplus a}(9) = 0.008, \quad \mu_{a\boxplus a\boxplus a}(x) = 0 \qquad \text{for other } x \in R.$$

Obviously, the vanishing of the marginal values illustrated by the example does not happen if the possibilities of the marginal values of the inputs are equal to 1. In the notation of (8.10) and (8.11), if $\mu_a(x_1) = \mu_a(x_2) = 1.0$.

The previous example illustrates another pleasant property of the convolutive procedures. They reflect, much better than do the supremum-minimum ones, the eventual asymmetry of the input values — not only by the asymmetry of the margins but also by shifting the top possibilities to values closer to the sums of the more possible inputs.

It is also useful to note the close relation of the convolutive representation to the probabilistic models of uncertainty. This identical processing of both probabilistic and fuzzy types of uncertainty can be understood as an attempt to ignore the differences between those types of uncertainty. In fact, the variety of vagueness modeled by fuzzy quantities is so wide that some fuzzy theoretical fundamentalism is not adequate for it. Different types of applied algorithms dealing with fuzzy data will probably demand different types of the representation principle, respecting their inner structure and specific features.

9

$\mathcal{L}$-Fuzzy Quantities

In the remainder of this book we keep Zadeh's classical model of fuzziness represented by the membership functions with values in the $[0,1]$ interval. It is very manageable and in this sense also more applicable in practical computing algorithms.

However, there exist situations in which the memberships determined by exact real numbers from $[0,1]$ do not fully correspond to the complexity and depth of uncertainty contained in the modeled reality. For this purpose some generalizations of the original $[0,1]$-memberships were suggested. Most of them can be covered by the concept of so called *$\mathcal{L}$-fuzzy sets*, set forth, e.g., in [36], [86], and [28]. According to this model, the values of the membership functions are elements of a general lattice. Then the common $[0,1]$-fuzzy sets become a special case of general $\mathcal{L}$-fuzziness, since the $[0,1]$ interval with the infinum (minimum) and supremum (maximum) operations, and 0 and 1 as minimal and maximal elements, respectively, is also a lattice. Of course, $\mathcal{L}$-fuzzy subsets of R can be, due to [36], treated as $\mathcal{L}$-fuzzy quantities.

The $\mathcal{L}$-fuzzy sets and related $\mathcal{L}$-fuzzy concepts are especially useful for constructing wide models general enough to cover a large scale of particular special cases. The applicability of $\mathcal{L}$-fuzziness to solving concrete practical problems or preparing real software algorithms is rather limited. This is why we mention the theory of $\mathcal{L}$-fuzzy quantities only in this chapter, and very briefly.

9.1 $\mathcal{L}$-Fuzzy Sets

The concept of $\mathcal{L}$-fuzzy sets is based on abstract lattice structures. A *lattice* can be represented by a quintuple

$$\mathcal{L} = (L, \wedge, \vee, \mathbf{0}, \mathbf{1}) \tag{9.1}$$

where L is a general non-empty set, $\wedge$ and $\vee$ are *binary operations* fulfilling the conditions of

$$
\begin{aligned}
\textit{commutativity:}\ & x \wedge y = y \wedge x, \quad x \vee y = y \vee x, \\
\textit{associativity:}\ & x \wedge (y \wedge z) = (x \wedge y) \wedge z, \quad x \vee (y \vee z) = (x \vee y) \vee z, \\
\textit{absorption:}\ & (x \vee y) \wedge x = x, \quad (x \wedge y) \vee x = x
\end{aligned}
$$

for all x, y, $z \in L$. $\mathbf{0}$ and $\mathbf{1}$ are the *minimal* and *maximal* elements of L (respectively), i.e., such that

$$
\begin{aligned}
x \wedge \mathbf{1} = x, &\quad x \vee \mathbf{1} = \mathbf{1}, \\
x \wedge \mathbf{0} = \mathbf{0}, &\quad x \vee \mathbf{0} = x,
\end{aligned}
$$

for any $x \in L$.

The properties of $\wedge$ and $\vee$ and the characteristics of $\mathbf{0}$ and $\mathbf{1}$ can be used in a more general sense to define the ordering relation over L by

$$x \leq y \quad \text{iff} \quad x \wedge y = x \tag{9.2}$$

which is equivalent to

$$x \leq y \quad \text{iff} \quad x \vee y = y \tag{9.3}$$

It is sometimes (not always) supposed that the lattice of the memberships in the fuzzy set theory is also *complete*, i.e., for any $K \subset L$ there exist

$$\bigwedge_{x \in K} x \quad \text{and} \quad \bigvee_{x \in K} x \tag{9.4}$$

in L, and that it is *distributive*, i.e.,

$$
\begin{aligned}
(x \vee y) \wedge z = (x \wedge z) \vee (y \wedge z), \\
(x \wedge y) \vee z = (x \vee z) \wedge (y \vee z),
\end{aligned}
$$

for any x, y, $z \in L$, or even that it is *infinitely distributive:*

$$
\begin{aligned}
x \wedge \left(\bigvee_{y \in K} y \right) &= \bigvee_{y \in K} (x \wedge y), \\
x \vee \left(\bigwedge_{y \in K} y \right) &= \bigwedge_{y \in K} (x \vee y).
\end{aligned}
$$

Each $\mathcal{L}$-*fuzzy subset* A of a universum $\mathcal{U}$ is defined by the mapping $\mu_A : \mathcal{U} \to \mathcal{L}$, which is called the membership function of A. Of course, analogously to the $[0,1]$-approach, for A and B being $\mathcal{L}$-fuzzy subsets of $\mathcal{U}$, their union and intersection are also $\mathcal{L}$-fuzzy sets, where for any $x \in \mathcal{U}$

$$\mu_{A\cup B}(x) = \mu_A(x) \vee \mu_B(x), \quad \mu_{A\cap B}(x) = \mu_A(x) \wedge \mu_B(x) \tag{9.5}$$

and

$$\mu_{\mathcal{U}}(x) = \mathbf{1}, \quad \mu_{\emptyset}(x) = \mathbf{0} \tag{9.6}$$

If A, B are fuzzy subsets of $\mathcal{U}$ then $A \subset B$ iff

$$\mu_A(x) \wedge \mu_B(x) = \mu_A(x) \quad \text{for all } x \in \mathcal{U} \tag{9.7}$$

This condition is, because of (9.2) and (9.3), equivalent to

$$\mu_A(x) \vee \mu_B(x) = \mu_B(x) \quad \text{for all } x \in \mathcal{U} \tag{9.8}$$

The concept of a complement of a fuzzy set can be defined as well, but this definition, demanding a few auxiliary notions, is not essential in this context.

9.2 $\mathcal{L}$-Fuzzy Quantities

We use the term *$\mathcal{L}$-fuzzy quantity* for any $\mathcal{L}$-fuzzy subset of the real line R fulfilling the conditions

$$\exists\, x_0 \in R : \; \mu_a(x_0) = \mathbf{1},$$
$$\exists\, x_1, x_2 \in R, \quad x_1 < x_1 : \; \mu_a(x) = \mathbf{0} \quad \text{for all } x \notin [x_1, x_2],$$

which are $\mathcal{L}$-fuzzy set theoretical analogies of (2.1) and (2.2). The set of all $\mathcal{L}$-fuzzy quantities will be denoted by $\mathbb{R}_{\mathcal{L}}$.

The representation principle can be formulated for $\mathcal{L}$-fuzzy quantities (see [36]) in the following way. If $f : R \times R \to R$ is a real-valued function of two arguments, then it can be extended to a mapping $f : \mathbb{R}_{\mathcal{L}} \times \mathbb{R}_{\mathcal{L}} \to \mathbb{R}_{\mathcal{L}}$, so that for $c = f(a, b)$, a, b, $c \in \mathbb{R}_{\mathcal{L}}$,

$$\mu_c(z) = \bigvee_{\substack{x, y \in R \\ z = f(x,y)}} (\mu_a(x) \wedge \mu_b(y)) \tag{9.9}$$

which is fully analogous to (2.13).

For the operations of addition and fuzzy product, the general representation theorem can be specified analogously to (3.1) and (3.7) as follows.

If $a,\ b \in \mathbb{R}_{\mathcal{L}}$ then the $\mathcal{L}$-fuzzy quantity $a \oplus b \in \mathbb{R}_{\mathcal{L}}$ with the membership function

$$\mu_{a\oplus b}(x) = \bigvee_{y\in R} (\mu_a(y) \wedge \mu_b(x-y)) \tag{9.10}$$
$$= \bigvee_{z\in R} (\mu_a(x-z) \wedge \mu_b(z))$$

is called the sum of a and b. It is easy to verify that $a \oplus b$, defined by (9.10), really belongs to $\mathbb{R}_{\mathcal{L}}$. If $z_0 \in R$ and $y_0 \in R$ are numbers fulfilling $\mu_a(y_0) = \mathbf{1}$ and $\mu_b(z_0) = \mathbf{1}$, then evidently $x_0 = y_0 + z_0$ fulfills $\mu_{a\oplus b}(x_0) = \mathbf{1}$. Analogously, if $z_1,\ z_2 \in R$ and $y_1,\ y_2 \in R$ fulfill

$$\mu_a(y) = \mathbf{0} \quad \text{for} \quad y \notin [y_1, y_2], \qquad \mu_b(z) = \mathbf{0} \quad \text{for} \quad z \notin [z_1, z_2]$$

then evidently $y_1 + z_1$ and $y_2 + z_2$ are real numbers for which

$$\mu_{a\oplus b}(x) = \mathbf{0} \quad \text{for} \quad x \notin [y_1 + z_1,\ y_2 + z_2]$$

Other properties of the addition, presented for the $[0,1]$-fuzzy quantities in Chapters 3 and 5 can also be analogously derived for the $\mathcal{L}$-fuzzy quantities. It means that the operation $\oplus$ is commutative and associative (see Equations 3.4 and 3.5) and that for $\langle r\rangle,\ r \in R$, defined by

$$\mu_{\langle r\rangle}(r) = \mathbf{1}, \quad \mu_{\langle r\rangle}(x) = \mathbf{0} \quad \text{for } x \neq r \tag{9.11}$$

and $a \in \mathbb{R}_{\mathcal{L}}$ equality (3.6) where $r = 0$ holds, i.e., $a \oplus \langle 0\rangle = a$. The concept of symmetry and the related notion of additive equivalence can also be dealt with analogously to sections 4.1 and 4.2.

The definition of the crisp product $r \cdot a,\ r \in R,\ a \in \mathbb{R}_{\mathcal{L}}$, given by (3.12) is unchanged, following immediately from its structure. The distributivity of the crisp product and $\oplus$ in the form

$$r \cdot (a \oplus b) = (r \cdot a) \oplus (r \cdot b), \quad r \in R,\ a,\ b \in \mathbb{R}_{\mathcal{L}} \tag{9.12}$$

is also fulfilled. On the other hand, the distributivity of the lattice $\mathcal{L}$ does not imply the validity of the complementary distributivity law $(r_1+r_2)\cdot a = r_1\cdot a \oplus r_2\cdot a$, as is illustrated by Example 36 in the following section.

An analogous situation exists for the fuzzy product operation $\odot$. Analogously to (3.7) and (3.8), we define for $a,\ b \in \mathbb{R}_{\mathcal{L}}$ the $\mathcal{L}$-fuzzy quantity $a \odot b \in \mathbb{R}_{\mathcal{L}}$ by

$$\mu_{a\odot b}(x) = \bigvee_{\substack{y\in R\\ y\neq 0}} (\mu_a(x/y) \wedge \mu_b(y)) \tag{9.13}$$

for any $x \in R_0$,

$$\mu_{a\odot b}(0) = \mu_a(0) \vee \mu_b(0) \tag{9.14}$$

The commutativity and associativity of the fuzzy product of $\mathcal{L}$-fuzzy quantities, analogous to (3.9) and (3.10), as well as the equality

$$a \odot \langle 1 \rangle = a, \quad a \in \mathbb{R}_{\mathcal{L}} \tag{9.15}$$

analogous to (3.11), can easily be derived from (9.13) and (9.14). Evidently, the concepts of transversibility and multiplicative equivalence can also be dealt with analogously to the case of $[0,1]$-fuzzy quantities.

9.3 Special Types of $\mathcal{L}$-Fuzzy Quantities

The generality of the concept of $\mathcal{L}$-fuzziness allows the suggestion of various special types of the general model.

The first is the classical $[0,1]$-*fuzzy set* model, treated in all other chapters of this book. Here the lattice quintuple (9.1) consists of the basic set $[0,1]$, operations of infinum (minimum) and supremum (maximum) representing $\wedge$ and $\vee$, respectively, and with the maximal $\mathbf{1}$ and minimal $\mathbf{0}$ elements 1 and 0, respectively. This lattice is complete but not distributive, and the properties of $\mathcal{L}$-fuzzy quantities of that type are investigated throughout this book.

Also the $[0,1]$-fuzzy set with operations of multiplication

$$x \cdot y \quad \text{instead of} \quad \wedge, \quad x,\, y \in [0,1]$$

addition with subtracted product,

$$x + y - x \cdot y \quad \text{instead of} \quad \vee, \quad x,\, y \in [0,1]$$

and maximal and minimal elements 1 and 0 form another special type of $\mathcal{L}$-fuzzy sets. Even this lattice is not distributive.

Another type of $\mathcal{L}$-fuzzy set and quantities is based on the lattice $\mathcal{L}$, where the set L is a class of all subsets of some closed compact set M, denoted by 2^M, where operations $\vee$ and $\wedge$ are the union $\cup$ and intersection $\cap$ in the usual set theoretical sense, respectively, the maximal element $\mathbf{1}$ is the whole set M, and the minimal element $\mathbf{0}$ is the empty set $\emptyset$. We call this type of $\mathcal{L}$-fuzzy set a *set-based $\mathcal{L}$-fuzzy set.* This lattice is complete and also distributive, or even infinitely distributive. However, the arithmetic operations of addition and crisp product used for the related $\mathcal{L}$-fuzzy quantities from $\mathbb{R}_{\mathcal{L}}$ are not distributive, i.e., for $r_1, r_2 \in R$, $a \in \mathbb{R}_{\mathcal{L}}$, the equality $(r_1 + r_2) \cdot a = r_1 \cdot a \oplus r_2 \cdot a$ does not generally hold. This is shown in the next example, which in this sense illustrates one of the general statements presented in the previous section.

Example 36. Let us consider a set-based $\mathcal{L}$-fuzzy quantity $a \in \mathbb{R}_{\mathcal{L}}$, where $M = [0, 1]$,

$$\mathcal{L}\left(2^{[0,1]}, \cap, \cup, \emptyset, [0,1]\right)$$

This means that the values $\mu_a(x) \subset [0, 1]$ for any $x \in R$. Let the values of μ_a be

$$\mu_a(1) = [0.5,\ 0.75],\ \ \mu_a(2) = [0, 1],\ \ \mu_a(4) = (0,\ 0.6),\ \ \mu_a(x) = \emptyset$$

for other $x \in R$.

Let $r_1 = r_2 = 1$. Then

$$(r_1 + r_2) \cdot a = 2 \cdot a, \quad \text{and}$$
$$\mu_{2a}(2) = [0.5,\ 0.75], \quad \mu_{2a}(4) = [0, 1], \quad \mu_{2a}(8) = (0,\ 0.6),$$
$$\mu_{2a}(x) = \emptyset \quad \text{for other } x \in R.$$

On the other hand, using (9.9),

$$(r_1 \cdot a) \oplus (r_2 \cdot a) = a \oplus a, \quad \text{and}$$
$$\mu_{a\oplus a}(2) = [0.5,\ 0.75], \quad \mu_{a\oplus a}(3) = [0.5,\ 0.75],$$
$$\mu_{a\oplus a}(4) = [0, 1], \quad \mu_{a\oplus a}(5) = [0.5,\ 0.6), \quad \mu_{a\oplus a}(6) = (0,\ 0.6),$$
$$\mu_{a\oplus a}(8) = (0,\ 0.6), \quad \mu_{a\oplus a}(x) = \emptyset \quad \text{for other } x \in R.$$

This means that in this case

$$(r_1 + r_2) \cdot a \neq (r_1 \cdot a) \oplus (r_2 \cdot a)$$

In this example the set L was a class of subsets of a closed interval. Of course, it can even be a class of much more general sets, including sets of real objects, events, etc., if it corresponds with the specific demands of the modeled situation.

The last special type of $\mathcal{L}$-fuzzy sets and quantities mentioned here are so-called *ultra-fuzzy sets*. Their construction was motivated by certain paradoxes connected with the concept of fuzziness. In the case of a traditional $[0, 1]$-fuzzy set we, for a given $x \in \mathcal{U}$ and $A \subset_f \mathcal{U}$, do not certainly know if $x \in A$, but we pretend to know very exactly the degree of that uncertainty, as represented by the value of the membership function $\mu_a(x) \in [0, 1]$. The idea of the ultra-fuzzy set was an attempt to overcome this paradox by shifting it one step further. An ultra-fuzzy set is, e.g., in [111], an $\mathcal{L}$-fuzzy set where L is the class of fuzzy subsets of the interval $[0, 1]$. The operations $\vee$ and $\wedge$ are then the union and intersection of fuzzy sets, defined as a consequence of (1.2) and (1.3).

(Do not forget that the operations of minimum and maximum in this case concern the values of membership functions of some classical [0, 1]-fuzzy sets, which themselves are generalized values of the memberships of some $\mathcal{L}$-fuzzy set.) If we denote by $A \subset_f \mathcal{U}$ an ultra-fuzzy subset of some universum $\mathcal{U}$ and by $\mu_A(x)$ the value of its membership function for some $x \in \mathcal{U}$, then we denote by

$$m_{A,x} : [0,1] \to [0,1]$$

the membership function of the fuzzy subset $\mu_A(x) \subset_f [0,1]$ of the unit interval. The maximal element of the lattice $\mathcal{L}$ is, in this case, the fuzzy subset **1** of [0, 1], such that

$$\mu_A(x) = \mathbf{1} \quad \text{means} \quad m_{A,x}(\xi) = 1 \quad \text{for all } \xi \in [0,1] \tag{9.16}$$

and the minimal element is the fuzzy subset **0** of [0, 1] such that

$$\mu_A(x) = \mathbf{0} \quad \text{means} \quad m_{A,x}(\xi) = 0 \quad \text{for all } \xi \in [0,1] \tag{9.17}$$

The lattice of the membership values of ultra-fuzzy sets is complete but not generally distributive.

The ultra-fuzzy quantity is an ultra-fuzzy subset of the real line. Each such fuzzy quantity $a \in \mathbb{R}_{\mathcal{L}}$ is defined by a class of membership functions

$$\{m_{a,x} : x \in R,\ m_{a,x} : [0,1] \to [0,1]\} \tag{9.18}$$

with properties corresponding to the general properties of $\mathcal{L}$-fuzzy quantities. Their formal description is unfortunately rather complicated due to the complex structure of the input elements. For example, the sum of two ultra-fuzzy quantities a and b is an ultra-fuzzy quantity $a \oplus b \in \mathbb{R}$, where, with respect to (9.10) and (9.18),

$$\mu_{a\oplus b}(x) = \bigvee_{y\in R} (\mu_a(y) \wedge \mu_b(x-y)) \quad \text{for all } x \in R$$

i.e.,

$$m_{a\oplus b,x}(\xi) = \sup_{y\in R} (\min(m_{a,y}(\xi),\, m_{b,x-y}(\xi)))$$

for all $\xi \in [0,1]$ and $x \in R$.

Similarly, if $a \in \mathbb{R}_{\mathcal{L}}$, $r \in R$ then for $r \cdot a \in \mathbb{R}_{\mathcal{L}}$ and all $x \in R$

$$\begin{aligned} \mu_{r\cdot a}(x) &= \mu_a(x/r) \quad \text{for } r \neq 0, \\ &= \mu_{\langle 0\rangle}(x) \quad \text{for } r = 0, \end{aligned}$$

which means that for all $x \in R$ and all $\xi \in [0,1]$

$$\begin{aligned} m_{r\cdot a,x}(\xi) &= m_{a,x/r}(\xi), \qquad \text{if } r \neq 0, \\ &\left.\begin{aligned} &= 1 \quad \text{for } x = 0, \\ &= 0 \quad \text{for } x \neq 0 \end{aligned}\right\} \text{if } r = 0 \end{aligned}$$

In this case we have used (2.4), (9.16), and (9.17), which implies that for $r \in R$ the ultra-fuzzy quantity $\langle r \rangle \in \mathbb{R}_{\mathcal{L}}$ fulfills

$$\mu_{\langle r \rangle}(r) = \mathbf{1}, \quad \mu_{\langle r \rangle}(x) = \mathbf{0} \quad \text{for } x \neq r$$

i.e., for all $\xi \in [0, 1]$

$$m_{\langle r \rangle, r}(\xi) = 1, \quad m_{r,x}(\xi) = 0 \quad \text{if } x \neq r$$

Part IV

Applications

10

Fuzzy Data

The concept of a fuzzy set and the other concepts derived from it, including fuzzy quantities, were suggested to offer a mathematical tool for modeling the vagueness of reality. This means that vague numeric data belong to one of the potential and important application areas of the fuzzy quantities theory. The standard mathematical models of real data are usually constructed as a mixture of some unknown deterministic "true" value and some kind of noise contaminating that value. The characteristic of the noise is sometimes supposed to be symmetric in some sense. Here we suppose, of course, that the noise is vaguely described and characterized by a fuzzy quantity.

10.1 Structure of Contaminated Data

In keeping with the general heuristic description of real data presented above, we suppose that fuzzy data are a mixture of some crisp value $x \in R$ (or the corresponding degenerated fuzzy quantity $\langle x \rangle \in \mathbb{R}$) and fuzzy noise represented by a fuzzy quantity $a \in \mathbb{R}$. Of course it is possible to suppose multidimensional fuzzy data, using the results of Section 6.2.

The concept of fuzziness should not replace other (probabilistic) models in cases in which they are adequate for the situation. At any rate, there exist many real problems in which the data are not contaminated by a stochastic uncertainty but by some vagueness of a fuzzy type. This concerns, for example, the subjective but neither probabilistic nor statistical estimation of some numeric quantities, the non-stochastic imprecision of measurements or perceptors, verbally given values, etc. Such fuzzy data are usually given by statements such as: "a value not far from 100", "somewhere between 50 and 60, maybe in the upper part of the interval", "perhaps almost 70", but also "more than was expected", "a very small value", etc. Such statements are rarely supported by experience or past data rich enough to complete a responsible statistical estimation of their meaning. On the other hand, they contain some kind of evaluation of the obtained quantitative or qualitative information, allowing the construction of—perhaps only on a subjective basis—some membership function describing the possible values of the referenced numerical data, including even the supposed degree of their possibility.

There are usually two types of mixture of the deterministic and indeterministic components of fuzzy data considered in the literature — the additive and the multiplicative ones. More complicated combinations, however possible, are not usually taken into account. In our fuzzy case we keep the limitation too. This means that the crisp components $\langle x \rangle$ and the fuzzy components $a \in \mathbb{R}$ form the fuzzy datum $d \in \mathbb{R}$ either by addition

$$d = \langle x \rangle \oplus a \tag{10.1}$$

or by multiplication

$$d = \langle x \rangle \odot a = x \cdot a \tag{10.2}$$

Rather specific but frequently considered types of fuzzy data are those with a somehow symmetric noisy component. In the additive case this means that $a \in \mathbb{S}_0$ and, consequently, $d \in \mathbb{S}_x$. In the multiplicative case the symmetry naturally turns into transversibility and, consequently, $a \in \mathbb{T}_1,\ d \in \mathbb{T}_x$ as follows from (10.2).

10.2 Additive Noise

The general rules mentioned in Sections 3.1, 5.1, and 5.2 are valid even for fuzzy data of the additive type (10.1). Generally, if

$$d_1 = \langle x \rangle \oplus a, \qquad d_2 = \langle y \rangle \oplus b$$

are additive fuzzy data, then $d_1 \oplus d_2$ and $r \cdot d_1$, for $r \in R$, are also fuzzy data, and

$$d_1 \oplus d_2 = \langle x + y \rangle \oplus (a \oplus b), \qquad r \cdot d_1 = \langle r \cdot x \rangle \oplus (r \cdot a),$$

as follows from the associativity, commutativity, and distributivity of fuzzy quantities.

If $d_1 = \langle x \rangle \oplus a,\ d_2 = \langle y \rangle \oplus (-a)$, then, according to the results of Section 5.1, $d_1 \oplus d_2 \sim_{\oplus} \langle x + y \rangle$, which means that $d_1 \oplus d_2 = \langle x + y \rangle \oplus s$ for some $s \in \mathbb{S}_0$, or, symbolically, $d_1 \oplus d_2 \in \mathbb{S}_{x+y}$.

As a result of Section 5.2, data with symmetric additive noise form a linear space with additive equivalence $\sim_{\oplus}$ instead of an equality. If

$d = \langle x \rangle \oplus a$ and $a \in \mathbb{S}_y$ for some $y \in R$, then $d = \langle x + y \rangle \oplus s$ for $s \in \mathbb{S}_0$ such that $a = \langle y \rangle \oplus s$ or $s = \langle -y \rangle \oplus a$.

10.3 Multiplicative Noise

Even in the multiplicative case the general rules mentioned in Sections 3.2 and 5.3 represent basic tools for computations with fuzzy data. They imply that for data $d_1 = \langle x \rangle \odot a$, $d_2 = \langle y \rangle \odot b$ and for $r \in R$ the products $d_1 \odot d_2$ and $r \cdot d_1$ have the form

$$d_1 \odot d_2 = \langle x \cdot y \rangle \odot (a \odot b), \qquad r \cdot d_1 = \langle r \cdot x \rangle \odot a$$

If $d_1 = \langle x \rangle \odot a$, $d_2 = \langle x \rangle \odot b$ then the sum $d_1 \oplus d_2$ also has the structure of fuzzy data with multiplicative noise, namely $d_1 \oplus d_2 = \langle x \rangle \odot (a \oplus b)$, as follows from the distributivity rule (3.14) and from (3.13).

The character of multiplicative noise contaminating the fuzzy data means that it is natural to suppose its positivity, i.e., its belonging to the set $\mathbb{R}^+$. The eventual negative noise from $\mathbb{R}^-$ can be transformed into positive noise by changing the sign of the crisp component of the data. Since the noise is supposed to be positive, the fuzzy data with multiplicative noise are signed fuzzy quantities from $\mathbb{R}^*$, and the results summarized in Section 5.3 can be applied.

If, moreover, the multiplicative noise is y-transversible for some $y > 0$, i.e., if $d = \langle x \rangle \odot a$, $a \in \mathbb{T}_y$ then $d = \langle x \cdot y \rangle \odot t$ for $t \in \mathbb{T}_1$ such that $a = \langle y \rangle \odot t$ or $t = (1/y) \cdot a$.

If, on the other hand, $a \in \mathbb{S}_y \cap \mathbb{R}^+$ for some $y > 0$, then the data with multiplicative noise $d = \langle x \rangle \odot a$ are also data with additive noise, namely

$$d = \langle x \cdot y \rangle \oplus s$$

where $s \in \mathbb{S}_0$ is the 0-symmetric fuzzy quantity, for which $s = x \cdot s_0$, where $a = \langle y \rangle \oplus s_0$, $s_0 \in \mathbb{S}_0$. This means that for $a \in \mathbb{S}_y$, $d = \langle x \rangle \odot a \sim_\oplus \langle x \cdot y \rangle$.

11

Fuzzy Critical Path Method

The critical path method of network analysis was developed to optimize the realization of complex production or service processes consisting of many parallel or subsequent particular activities. In some cases the method is used to analyze qualitatively new or essentially modified procedures whose components are known only vaguely. It concerns especially the time duration of activities which can be, in many cases, only given by subjective estimation or even by vague verbal information. Some comments on this subject are presented in [60], [62], and [61].

11.1 Network Analysis

Network analysis and its substituent, the Critical Path Method (CPM), are based on the concept of a node-oriented network, which is defined as a quadruple

$$(n,\ (A_0, A_1, \ldots, A_n, A_{n+1}),\ (\mathcal{B}_1, \ldots, \mathcal{B}_{n+1}),\ (t_0, t_1, \ldots, t_n, t_{n+1})) \tag{11.1}$$

describing the production or construction process being considered. In (11.1) n is the number of *actual activities* existing in the procedure, A_i, $i = 1, \ldots, n$, are the activities themselves, and A_0 and A_n are, respectively, dummy *starting* and *terminal* activities, $t_i \in R$, $t_i \geq 0$, $i = 0, 1, \ldots, n, n+1$, and the *durations* of the respective activities, where by definition $t_0 = t_{n+1} = 0$. Finally $\mathcal{B}_i$, $i = 1, \ldots, n, n+1$ are sets of activities, $\mathcal{B}_i \subset \{A_0, A_1, \ldots, A_n, A_{n+1}\}$, defining their sequence. Each $\mathcal{B}_j$, $j = 1, \ldots, n+1$, is the set of activities *immediately preceding* activity A_j. We suppose that each A_i, $i = 1, \ldots, n$ belongs to some set $\mathcal{B}_j$ (and evidently $\mathcal{B}_0$, if considered, must be empty). Moreover, there does not exist a sequence of indices $(i_1, \ldots, i_m)$ such that for $j = 1, \ldots, m$ $A_{i_j} \in \mathcal{B}_{i_{j+1}}$ and $i_m = i_1$ (i.e., the sets $\mathcal{B}_i$ do not allow "cycles" of activities). If

$A_i \notin \mathcal{B}_j$ for all $j = 1, \ldots, n$, then $A_i \in \mathcal{B}_{n+1}$. If $\mathcal{B}_j \cap \{A_1, \ldots, A_n\} = \emptyset$, then $\mathcal{B}_j = \{A_0\}$, $j = 1, \ldots, n+1$, and $A_{n+1} \notin \mathcal{B}_j$ for any $j = 1, \ldots, n+1$.

Example 37. The previous notions can be illustrated by the following example of a network with 6 real activities

$$(6, (A_0, A_1, \ldots, A_6, A_7), (\mathcal{B}_1, \ldots, \mathcal{B}_6, \mathcal{B}_7), (0, t_1, \ldots, t_6, 0))$$

where

$$\mathcal{B}_1 = \{A_0\}, \mathcal{B}_2 = \mathcal{B}_3 = \{A_1\}, \mathcal{B}_4 = \mathcal{B}_5 = \{A_3\},$$
$$\mathcal{B}_6 = \{A_2, A_4\},\ \mathcal{B}_7 = \{A_5, A_6\}$$
$$t_1 = 2,\ t_2 = 4,\ t_3 = 3,\ t_4 = 2,\ t_5 = 6,\ t_6 = 1$$

This structure can be graphically represented by a node-oriented network (Figure 11.1), where nodes represent particular activities and the oriented arcs (arrows) the sequence of these activities, formally described by the sets $\mathcal{B}_i$.

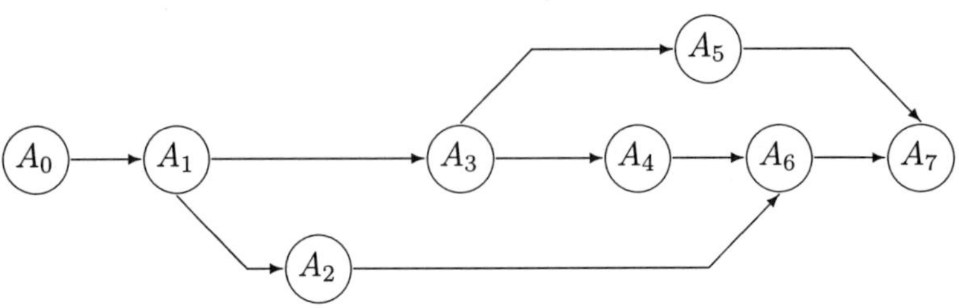

FIGURE 11.1

Every finite sequence of activities

$$P = \{A_{j_1}, A_{j_2}, \ldots, A_{j_m}\}, \qquad A_{j_k} \in \mathcal{B}_{j_{k+1}}, \qquad k = 1, 2, \ldots, m-1,$$
$$A_{j_1} = A_0, \qquad A_{j_m} = A_{n+1} \tag{11.2}$$

is called a *path*. If P is the path given above, then the number t_P defined by

$$t_P = t_{j_2} + \cdots + t_{j_{m-1}} \tag{11.3}$$

is called the *duration* of the path P (note that $t_{j_1} = t_0 = 0$ and $t_{j_m} = t_{n+1} = 0$). The set of all paths existing in the network is denoted by $\mathbf{P}$. The path $Q \in \mathbf{P}$ for which

$$t_Q \geq t_P \quad \text{for all } P \in \mathbf{P} \tag{11.4}$$

is the *critical path*, and the differences

$$r_P = t_Q - t_P \geq 0, \quad P \in \mathbf{P} \tag{11.5}$$

are called *floats* of the respective paths.

Example 38. If we consider the process described in the previous example, we see that there exist three paths

$$P = \{A_0, A_1, A_2, A_6, A_7\}, \qquad Q = \{A_0, A_1, A_3, A_5, A_7\},$$
$$S = \{A_0, A_1, A_3, A_4, A_6, A_7\}$$

with durations $t_P = 7$, $t_Q = 11$, and $t_S = 8$. Then the path Q is the critical path and the floats of the other paths are $r_P = 4$ and $r_S = 3$.

The analysis of such networks enables us to determine the critical path and the critical activities (forming the critical path), whose delay means an unavoidable delay of the whole controlled process. Moreover, the analysis of floats (which is well managed in network analysis methods) allows the determination of the existing floats of particular non-critical activities, i.e., the admissible delay in their realization, which cannot influence the total duration of the whole process.

11.2 Fuzzy Durations

The first network elements that can be fuzzified are the durations of activities, which are, in many non-standard production processes, apparently more or less vague. Such fuzzy durations are represented by fuzzy quantities from $\mathbb{R}$ either directly or in the form of fuzzy data decomposed into their deterministic and fuzzy noisy components (see Chapter 10). If we accept the representation of fuzzy durations by fuzzy data then additive noise seems to be more adequate to the additive character of network-analysis operations than does multiplicative noise.

If we represent the durations $t_1, \ldots, t_n$ by fuzzy quantities from $\mathbb{R}$, then it is natural to suppose that they are non-negative, i.e.,

$$\mu_{t_i}(x) = 0 \quad \text{for } x < 0,\ i = 1, \ldots, n \tag{11.6}$$

According to the interpretation given above, $t_0 = t_{n+1} = \langle 0 \rangle$. Then for any path $P = \{A_{j_1}, \ldots, A_{j_m}\}$ the duration t_P is also a fuzzy quantity fulfilling (11.6) and defined by the sum

$$t_P = t_{j_1} \oplus t_{j_2} \oplus \cdots \oplus t_{j_{m-1}} \oplus t_{j_m} = t_{j_2} \oplus \cdots \oplus t_{j_{m-1}} \tag{11.7}$$

If we represent the fuzzy durations of activities by fuzzy data such as in (10.1), i.e.,

$$t_i = \langle x_i \rangle \oplus u_i, \quad u_i \in \mathbb{R}^+,\ x_i \geq 0,\ i = 1, \ldots, n$$

then the sum (11.7) turns into

$$t_P = \left\langle x_{j_2} + \cdots + x_{j_{m-1}} \right\rangle \oplus \left(u_{j_2} \oplus \cdots \oplus u_{j_{m-1}} \right) \tag{11.8}$$

which preserves the standard form of fuzzy data with additive noise.

Example 39. Let us consider a network with the same activities and structure as Examples 37 and 38 but with fuzzy durations of activities. Let us denote by $\mu_i : R \to [0,1]$ the membership function of t_i, $i = 1, \ldots, n$, and let $\mu_i(x) = 0$ for $i = 1, \ldots, 6$, $x \in R$, except for the values

$$\begin{aligned}
&\mu_1(1) = 0.2, \quad \mu_1(2) = 1.0, \quad \mu_1(3) = 0.1,\\
&\mu_2(4) = 1.0, \quad \mu_2(5) = 0.2,\\
&\mu_3(3) = 1.0,\\
&\mu_4(3) = 0.5, \quad \mu_4(4) = 1.0,\\
&\mu_5(5) = 0.1, \quad \mu_5(6) = 1.0, \quad \mu_5(7) = 0.3,\\
&\mu_6(1) = 1.0, \quad \mu_6(2) = 0.2.
\end{aligned}$$

Then the durations of the paths P, Q, and S are fuzzy quantities with membership functions μ_P, μ_Q, and μ_S, respectively, where

$$\begin{gathered}
\mu_P(6) = 0.2,\ \mu_P(7) = 1.0,\ \mu_P(8) = 0.2,\ \mu_P(9) = 0.2,\ \mu_P(10) = 0.1,\\
\mu_Q(9) = 0.1,\ \mu_Q(10) = 0.2,\ \mu_Q(11) = 1.0,\ \mu_Q(12) = 0.3,\ \mu_Q(13) = 0.1,\\
\mu_S(8) = 0.2,\ \mu_S(9) = 0.5,\ \mu_S(10) = 1.0,\ \mu_S(11) = 0.2,\ \mu_S(12) = 0.1
\end{gathered}$$

and

$$\mu_P(x) = 0,\ \mu_Q(x) = 0,\ \mu_S(x) = 0, \text{ in the other cases.}$$

11.3 Fuzzy Critical Path and Floats

If the durations of paths are fuzzy quantities then the concept of the critical path also becomes fuzzy. Particular paths from **P** can be critical with a certain degree of possibility (depending on the possibilities of their durations), and the critical paths form a fuzzy subset of **P**.

Using the general methods presented in Section 2.2, we can construct the fuzzy set of critical paths in the following way. Let P, $S \in \mathbf{P}$ be paths with fuzzy durations t_P, $t_S \in \mathbb{R}$, where μ_P and μ_S are their membership functions. Then the possibility that the duration of P is not less than the duration of S is given by the number (see Equation 2.9)

$$\nu_{\succsim}(P, S) = \sup_{x,y \in R,\, x \geq y} (\min(\mu_P(x),\, \mu_S(y))) \tag{11.9}$$

where the relation "*duration of* $\cdot$ *is not less than the duration of* $\cdot$" is a fuzzy relation over $\mathbf{P}$.

If we denote by $\pi : \mathbf{P} \to [0, 1]$ the membership function of the fuzzy set of critical paths, then (see Equation 2.11 in Section 2.2) for $P \in \mathbf{P}$

$$\pi(P) = \min\left(\nu_{\succsim}(P, S) : S \in \mathbf{P}\right) \tag{11.10}$$

(which is the possibility that the duration of P is not less than the duration of any S in $\mathbf{P}$).

Since there is not merely one critical path (or a deterministic set of critical paths of equal durations) and since the durations of particular fuzzy critical paths are not crisp, the floats become rather more complex but also more informative objects.

If P, $S \in \mathbf{P}$ are paths then the difference of their durations

$$r_{SP} = t_S \oplus (-t_P) \in \mathbb{R} \tag{11.11}$$

is a fuzzy quantity representing the *relative float* of the path P with respect to the path S. In many cases the relative floats offer enough information for the analysis of risks connected with the time-realization of paths and activities. Sometimes it is useful to analyze more general floats related to all paths, with respect to the possibility that they might become critical. If we denote by $\mu_{SP} : R \to [0, 1]$ the membership function of the relative float r_{SP}, then we define the absolute float of the path P as a fuzzy quantity $\overline{r}_P \in \mathbb{R}$ with membership function $\rho_P : R \to [0, 1]$ given by

$$\rho_P(x) = \max_{S \in \mathbf{P}} (\min(\mu_{SP}(x),\, \pi(S))) \tag{11.12}$$

which is the possibility that the value $x \in R$ of the relative float r_{SP} is achieved for P with respect to at least one path S, and that the path S is critical.

Example 40. Let us consider again the network described in the previous example. Then using (11.9) it is possible to calculate the membership function $\nu(\cdot\,,\cdot)$ of the fuzzy relation "is not greater than" over the set $\mathbf{P}$, where

$$\nu_{\succsim}(S,P) = 1.0, \quad \nu_{\succsim}(P,S) = 0.2,$$
$$\nu_{\succsim}(P,Q) = 0.1, \quad \nu_{\succsim}(Q,P) = 1.0,$$
$$\nu_{\succsim}(Q,S) = 1.0, \quad \nu_{\succsim}(S,Q) = 0.2,$$

and, of course, $\nu_{\succsim}(P,P) = \nu_{\succsim}(Q,Q) = \nu_{\succsim}(S,S) = 1.0$. Then the membership function $\pi(\cdot)$ of the fuzzy set of critical paths achieves values

$$\pi(P) = 0.1, \quad \pi(Q) = 1.0, \quad \pi(S) = 0.2.$$

The calculation of relative floats using (11.11) leads to the following results.

$$\mu_{PQ}(-7) = 0.1, \quad \mu_{PQ}(-6) = 0.2, \quad \mu_{PQ}(-5) = 0.3, \quad \mu_{PQ}(-4) = 1.0,$$
$$\mu_{PQ}(-3) = 0.2, \quad \mu_{PQ}(-2) = 0.2, \quad \mu_{PQ}(-1) = 0.2, \quad \mu_{PQ}(0) = 0.1,$$
$$\mu_{PQ}(1) = 0.1, \quad \mu_{PQ}(x) = 0 \quad \text{for other } x \in R,$$

$$\mu_{PS}(-6) = 0.1, \quad \mu_{PS}(-5) = 0.2, \quad \mu_{PS}(-4) = 0.2, \quad \mu_{PS}(-3) = 1.0,$$
$$\mu_{PS}(-2) = 0.5, \quad \mu_{PS}(-1) = 0.2, \quad \mu_{PS}(0) = 0.2, \quad \mu_{PS}(1) = 0.2,$$
$$\mu_{PS}(2) = 0.1, \quad \mu_{PS}(x) = 0 \quad \text{for other } x \in R,$$

$$\mu_{QS}(-3) = 0.1, \quad \mu_{QS}(-2) = 0.1, \quad \mu_{QS}(-1) = 0.2, \quad \mu_{QS}(0) = 0.2,$$
$$\mu_{QS}(1) = 1.0, \quad \mu_{QS}(2) = 0.5, \quad \mu_{QS}(3) = 0.3, \quad \mu_{QS}(4) = 0.2,$$
$$\mu_{QS}(5) = 0.1, \quad \mu_{QS}(x) = 0 \quad \text{for other } x \in R.$$

Naturally, $\mu_{QP}(x) = \mu_{PQ}(-x)$, $\mu_{SP}(x) = \mu_{PS}(-x)$ and $\mu_{SQ}(x) = \mu_{QS}(-x)$ for all $x \in R$, as follows from (11.11). The same formula implies that $r_{PP} \in \mathbb{S}_0$, $r_{QQ} \in \mathbb{S}_0$ and $r_{SS} \in \mathbb{S}_0$ are 0-symmetric, and

$$\mu_{PP}(-4) = \mu_{PP}(4) = 0.1, \quad \mu_{PP}(-3) = \mu_{PP}(3) = 0.2,$$
$$\mu_{PP}(-2) = \mu_{PP}(2) = 0.2, \quad \mu_{PP}(-1) = \mu_{PP}(1) = 0.2,$$
$$\mu_{PP}(0) = 1.0,$$

$$\mu_{QQ}(-4) = \mu_{QQ}(4) = 0.1, \quad \mu_{QQ}(-3) = \mu_{QQ}(3) = 0.1,$$
$$\mu_{QQ}(-2) = \mu_{QQ}(2) = 0.2, \quad \mu_{QQ}(-1) = \mu_{QQ}(1) = 0.3,$$
$$\mu_{QQ}(0) = 1.0,$$

$$\mu_{SS}(-4) = \mu_{SS}(4) = 0.1, \quad \mu_{SS}(-3) = \mu_{SS}(3) = 0.2,$$
$$\mu_{SS}(-2) = \mu_{SS}(2) = 0.2, \quad \mu_{SS}(-1) = \mu_{SS}(1) = 0.5,$$
$$\mu_{SS}(0) = 1.0.$$

The absolute floats calculated using (11.12) are fuzzy quantities with membership functions

$$\begin{aligned}
\rho_P(x) &= 0.1 \quad \text{for } x = -4, -3, -2, 7, \\
&= 0.2 \quad \text{for } x = -1, 0, 1, 2, 3, 6, \\
&= 0.3 \quad \text{for } x = 5, \\
&= 1.0 \quad \text{for } x = 4, \\
&= 0 \qquad \text{for other } x \in R.
\end{aligned}$$

$$\begin{aligned}
\rho_Q(x) &= 0.1 \quad \text{for } x = -7, -6, -5, 3, 4, \\
&= 0.2 \quad \text{for } x = -4, -3, -2, 2, \\
&= 0.3 \quad \text{for } x = -1, 1, \\
&= 1.0 \quad \text{for } x = 0, \\
&= 0 \qquad \text{for other } x \in R.
\end{aligned}$$

$$\begin{aligned}
\rho_S(x) &= 0.1 \quad \text{for } x = -4, -3, -2, 7, \\
&= 0.2 \quad \text{for } x = -1, 0, 1, 2, 3, 6, \\
&= 0.3 \quad \text{for } x = 5, \\
&= 0.5 \quad \text{for } x = 4, \\
&= 1.0 \quad \text{for } x = 1, \\
&= 0 \qquad \text{for other } x \in R.
\end{aligned}$$

The method of processing fuzzy durations is worth a few comments. First, as mentioned in the preceding example, formula (11.11) immediately means that for any P, $S \in \mathbf{P}$

$$r_{SP}(x) = r_{PS}(-x) \quad \text{for any } x \in R$$

and

$$r_{PP}(x) = r_{PP}(-x) \quad \text{for any } x \in R, \text{ i.e., } r_{PP} \in \mathbb{S}_0$$

Further, the procedure described above implies that its steps are correct in the sense that the obtained fuzzy quantity r_{SP} belongs to $\mathbb{R}$, fulfilling (2.1) and (2.2). Moreover, (11.9) together with the fact that the durations of paths t_P fulfill (2.1) imply that there exists at least one path $P \in \mathbf{P}$ such that $\nu_{\succsim}(P, S) = 1$ for all $S \in \mathbf{P}$. This means also that $\pi(Q) = 1$ for at least one $Q \in \mathbf{P}$ and, consequently, $\overline{r}_P$ fulfills (2.1) and

(2.2) for all $P \in \mathbf{P}$. This implies that $\overline{r}_P$ is a correct fuzzy quantity from $\mathbb{R}$ as well.

The relative and absolute floats offer an interesting interpretation, which can be useful for the analysis of real networks with fuzzy inputs. This concerns, specifically, the negative values of floats, i.e., the cases when $\rho_P(x) > 0$ or $\mu_{SP}(x) > 0$ for some $x < 0$. In classical network analysis and CPM theory, such a situation cannot arise. In the fuzzy CPM, where the durations of activities and paths are vague and can achieve more possible values, some of their combinations can lead to negative floats with certain degree of possibility. A positive membership value $\mu_{SP}(x)$ for some negative x simply means that the duration of the path P can be longer than the duration of S, where x is the difference. Then the value

$$\max\left(\mu_{SP}(x) : x < 0\right)$$

is the possibility that P becomes "more critical" than S.

The addition and subtraction procedures applied during the calculation of t_P and r_{PS} are unavoidably connected with the extension of vagueness of the quantities in question. It follows from the nature of the addition operation $\oplus$ that the interval of possible values (the support of the membership function) is larger with every addition (or subtraction) of a quantity to the existing sum. This phenomenon is natural — the vagueness really is cumulative as more vague factors influence the final result. On the other hand, the possibility value of some marginal values of the resulting fuzzy quantities is usually not very large. If some negative values of floats are sufficiently significant, it is useful to compare the relative float r_{SP} of P with respect to S with the relative float r_{SS} and r_{PP} of S and P with respect to themselves. They are the "fuzzy zeros," i.e., elements of $\mathbb{S}_0$, and their extent shows the internal fuzziness of S and P entering into the fuzzy quantity r_{SP}.

11.4 Further Fuzzifications

The durations of activities are not the only vague components of some real production processes analyzed by networks and CPM. Even the existence of some activities and their ordering in the production process (this means the set of activities $(A_1, \ldots, A_n)$ and the class of the sets $(\mathcal{B}_1, \ldots, B_{n+1})$ can be fuzzified.

The fuzzification of the set of activities $(A_1, \ldots, A_n)$ can be managed in the following way. If $\alpha_i < 1$ is the possibility that the activity A_i, $i = 1, \ldots, n$, will not be realized at all, then it is equivalent to the possibility that the duration of A_i vanishes. In the symbols of the previous sections, we put $\mu_i(0) = \alpha_i$ to express the possibility that the activity A_i will be omitted.

The vague ordering of activities in the considered process cannot be modeled without (luckily only formal) difficulties. An attempt to do so is presented in [61]. It is based on the idea of enlarging the network of the investigated process into a few parallel subgraphs — each with one of the possible orderings of activities — evaluated by possibility weights reflecting the possibilities of particular orderings. Of course, the same activities included in different subgraphs should be formally renamed to avoid paradoxes in the network structure. This method is rather complicated and, worse, respecting the weights of particular branches implies that the fuzzy quantities describing some durations (of activities and paths) as well as some floats are not formally correct — namely, they do not fulfill the normality condition (2.1). In fact, this normality condition is important for the formal purity of the additive equivalence concept, and it can be easily omitted in the context of the methods mentioned here for processing fuzzy networks. In any event, the complexity of this approach to the fuzzy ordering of activities leads to the conclusion that the optimal strategy of analyzing such processes is to construct a few completely separate networks, to analyze them simultaneously by the CPM methods, and to evaluate and responsibly interpret the obtained results.

12

Multicriteria Decision-Making

A further example of applications in the area of fuzzy set theory can be seen in decision-making theory. We focus on its multicriteria branch, where the variety of approaches offers an opportunity to use various concepts of fuzzy set and fuzzy quantities theory (see [16], [46], [70], [83], [116], [107], [17], [113], and [101]). Before doing so, we briefly mention the position of fuzziness in monocriterion decision-making, which illustrates the specificity of fuzzy theoretical thinking.

12.1 Monocriterion Decision-Making

The most common model of a decision-making problem is the following. There exists a non-empty set of *decisions* D, and a non-empty set $\mathcal{C}$ of *consequences* of those decisions. The decision-maker is able to distinguish among the consequences by means of the ordering relation $\succsim$ over $\mathcal{C}$, which is called the *criterion*. The relation $\succsim$ is frequently represented by a *utility function* $u : \mathcal{C} \to R$ such that $u(c) \geq u(c')$ iff $c \succsim c'$ for $c, c' \in \mathcal{C}$. Here we briefly recall that classical decision-making theory divides decision-making problems into three main classes.

In the first class any decision $d \in D$ is deterministically connected with a consequence $c \in \mathcal{C}$ (or with a utility $u(c)$) by means of a resulting function $\rho : D \to \mathcal{C}$. The decision-maker obviously accepts the decision that is connected with the optimal consequence, or with the maximal utility. The seemingly clear conclusion *can* involve complicated analytical procedures for finding the very best consequence (mathematical branches such as linear or quadratic programming and others are of this sort) but for our degree of generality the situation is quite transparent. Those decision-making problems are called decision-making *under certainty*.

The second class of problems includes decision-making in which each decision $d \in D$ has a probability distribution over the set $\mathcal{C}$ (or over the set $\{x \in R : \exists c \in \mathcal{C}, x = u(c)\}$. If we denote by $\mathcal{P}$ the class of probability distributions over $\mathcal{C}$, then the resulting function $\rho : D \to \mathcal{P}$ characterizes such decision-making, called decision-making *under risk.* There are various methods for solving such problems (some are developed in stochastic programming theory), partially optimizing the expected value of utilities, partially using other methods.

The third class contains the so called decision-making problems *under uncertainty.* Here the resulting function $\rho : D \to 2^{\mathcal{C}}$ connects every decision with a (non-empty) subset of consequences. Usual solution methods (which also include such branches as strategic games) are based on the choice of proper representatives of every set $\rho(d) \subset \mathcal{C}$ and on accepting the decision that prefers the set $\rho(d)$ with the best representative. The methods that accept for such representatives the "worst", i.e., least favorable or of the smallest utility, elements of the sets $\rho(d)$ are often called "mini-max" or "maxi-min" ones.

The problem consists in the fact that many real decision-making problems are in some sense a mixture of risk and uncertainty. The information about the resulting function is not good enough to offer a probability distribution over the possible outcomes, but on the other hand, the sets of consequences expected for every decision are at least vaguely structured — some of their elements are more possible or more expected, others (even if not excluded) seem to be less possible. In our terminology, the expected consequences of every decision frequently form a fuzzy subset of the set $\mathcal{C}$. Here is a good place to apply fuzzy set theory and its methods to monocriterion decision-making.

In previous paragraphs we supposed that the preferences (utilities) are perfectly known. This assumption is also not universal, and vague preferences may appear. If the decision-maker's preferences are represented not by deterministic utilities but by fuzzy utility functions, then the outcomes of particular decisions are fuzzy subsets of the set R, which means fuzzy quantities (if they fulfill Equations (2.1) and (2.2)). Their ordering and maximizing methods were briefly mentioned in Section 2.2, and they can form the theoretical basis for further developed optimization procedures derived for particular decision-making problems. From the formal point-of-view, decision-making under fuzzy uncertainty (where the resulting function connects each decision with a fuzzy subset of $\mathcal{C}$ and the utility functions are deterministic) and decision-making under certainty with fuzzy preferences lead to identical theoretical models, using the same approach based on the ordering of fuzzy quantities.

To illustrate the wide scale of possible applications of fuzzy quantities theory to decision-making models, we may instead use multicriteria decision-making problems, which allow the use of multidimensional forms.

12.2 Multicriteria Approaches

Let us consider a decision-making problem with a set D of decisions, set $\mathcal{C}$ of consequences, a deterministic resulting function $\rho : D \to \mathcal{C}$, and with n criteria $\succsim_1, \ldots, \succsim_n$, $n \geq 2$, each of them an ordering relation over $\mathcal{C}$. Let each criterion $\succsim_i$, $i = 1, \ldots, n$, also be represented by a utility function $u_i : \mathcal{C} \to R$. The main goal of any multicriteria decision-making method is to construct some compromise criterion $\succsim$ over $\mathcal{C}$, represented also by a compromise utility function $u : \mathcal{C} \to R$, which respects demands of all partial criteria and combines them in a somehow balanced way. The single strict demand put on $\succsim$ and u is that $c \succsim c'$ or $u(c) \geq u(c')$ if $c \succsim_i c'$ or $u_i(c) \geq u_i(c')$ for all $i = 1, \ldots, n$. The variety of existing practical methods applicable to the solution of this problem usually respects one of the following three basic approaches.

In the *lexicographic approach* it is supposed that the criteria (utility functions) are ordered according to their importance — $\succsim_i$ is more important than $\succsim_{i+1}$ for $i = 1, \ldots, n-1$. If we denote $D^{(0)} = D$ then the sequence of sets can be constructed:

$$\begin{aligned} D^{(i)} &= \left\{ d \in D^{(i-1)} : \forall\, d' \in D^{(i-1)},\ \rho(d) \succsim_i \rho(d') \right\} \qquad (12.1) \\ &= \left\{ d \in D^{(i-1)} : u_i(\rho(d)) = \max(u_i(\rho(d') : d' \in D^{(i-1)}) \right\}, \\ & i = 1, \ldots, n \end{aligned}$$

Evidently $D \supset D^{(1)} \supset \cdots \supset D^{(n)}$ and if the maxima in (12.1) exist then also $D^{(n)} \neq \emptyset$. Then, of course, $D^{(n)}$ is the set of optimal decisions.

The second approach covers *vector optimization* methods. They are based on the domination relation $\succsim$ over $\mathcal{C}$ (which can, using ρ, be easily transformed into a relation over D) such that $c \succsim c'$ iff

$$\begin{aligned} & c \succsim_i c' && \text{for all } i = 1, \ldots, n \\ & \neg(c' \succsim_j c) && \text{for at least one } j \in \{1, \ldots, n\} \end{aligned} \qquad (12.2)$$

which is equivalent to

$$\begin{aligned} & u_i(c) \geq u_i(c') && \text{for all } i = 1, \ldots, n \\ & u_j(c) > u_j(c') && \text{for at least one } j \in \{1, \ldots, n\} \end{aligned} \qquad (12.3)$$

The extension of relation $\succsim$ to the set D is easy if we put

$$d \succsim d' \quad \text{iff} \quad \rho(d) \succsim \rho(d'), \quad d, d' \in D$$

The relation $\succsim$ is generally only a partial one, which means that there may exist a large set of maximal elements. This is called a *Pareto optimum* and is denoted by $P_{\mathcal{C}} \subset \mathcal{C}$ where

$$\begin{aligned} P_{\mathcal{C}} &= \{c \in \mathcal{C} : \text{ there does not exist } c' \in \mathcal{C} \text{ such that } c' \succsim c\} \\ &= \{c \in \mathcal{C} : \forall c' \in \mathcal{C} \text{ either } \exists i \in \{1, \ldots, n\} \ u_i(c) > u(c') \\ &\quad \text{or } \forall i = 1, \ldots, n, \ u_i(c) = u_i(c')\} \end{aligned} \tag{12.4}$$

Using the resulting function ρ it is also possible to define the Pareto optimum P_D of the set D, i.e., by

$$P_D = \{d \in D : \rho(d) \in P_{\mathcal{C}}\} \tag{12.5}$$

There exists an alternative "strong" domination concept represented by the relation $\succ$ over $\mathcal{C}$ where $c \succ c'$ iff for all $i = 1, \ldots, n$ $u_i(c) > u_i(c')$. This strong domination can also be used to define alternative Pareto optima $P^*_{\mathcal{C}}$ or P^*_D analogously to (12.4) by

$$P^*_{\mathcal{C}} = \{c \in \mathcal{C} : \forall c' \mathcal{C}, \ \exists i \in \{1, \ldots, n\}, \ u_i(c) \geq u_i(c')\}$$

and P^*_D is defined using (12.5) with $P^*_{\mathcal{C}}$ instead of $P_{\mathcal{C}}$. This modified domination is really stronger, since $c \succ c'$ obviously implies $c \succsim c'$ and $P^*_{\mathcal{C}} \supset P_{\mathcal{C}}$. In the following sections, in accordance with majority of other papers, we prefer the relation $\succsim$ given in (12.2) and related Pareto optimum $P_{\mathcal{C}}$ defined by (12.4).

The last approach to multicriteria decision-making uses a combined *global criterion.* For its description the utility functions are essential. Generally, the global criterion is represented by a global utility function $u : \mathcal{C} \to R$ defined by a function $f : R^n \to R$

$$u(c) = f(u_1(c), \ldots, u_n(c)), \quad c \in \mathcal{C} \tag{12.6}$$

which is not decreasing in all variables. Very often the function f represents a convex combination

$$u(c) = w_1 \cdot u_1(c) + \cdots + w_n \cdot u_n(c), \quad c \in \mathcal{C} \tag{12.7}$$

where $w_1, \ldots, w_n$ are real-valued weights fulfilling

$$w_i \geq 0, \quad i = 1, \ldots, n, \qquad w_1 + \cdots + w_n = 1 \tag{12.8}$$

All three approaches described here as methods of finding maximal elements of the set $\mathcal{C}$ can be easily reformulated for finding the optimal decisions. It is sufficient to use the resulting function ρ and to define the utilities of decisions according to particular criteria by

$$u_i(d) = u_i(\rho(d)), \quad i = 1, \ldots, n, \ d \in D \tag{12.9}$$

The fuzzification of the multicriteria decision-making problem is based on the fuzzification of criteria. As with the monocriterial case, so with multicriteria decision-making in many practical situations — the criteria used by the decision-maker are in some degree vague. This concerns the orderings $\succsim_i$ as well as the corresponding utility functions u_i. In the following sections we assume that there exist utility functions u_i, $i = 1, \ldots, n$, of particular criteria, and that the vagueness of those criteria is formally represented by the fuzziness of functions u_i. As mentioned in Section 1.3, this is equivalent to the assumption that each value $u_i(c)$, $c \in \mathcal{C}$ (or $u_i(d)$, $d \in D$ in the sense of (12.9)) is a fuzzy subset of R or, in our terminology, that $u_i(c)$ or $u_i(d)$ are fuzzy quantities from $\mathbb{R}$ with membership functions $\mu_{i,c} : R \to [0,1]$ (or $\mu_{i,d} : R \to [0,1]$).

Let us now consider the approaches mentioned above from the point of view of the fuzziness of $u_i(c)$, $c \in \mathcal{C}$, $i = 1, \ldots, n$, as was done, e.g., in [72].

12.3 Fuzzy Lexicography

To construct fuzzy modifications of sets $D^{(1)}, D^{(2)}, \ldots, D^{(n)}$ described by (12.1), we need to use the method mentioned in Section 2.2 (and applied already in Section 11.3). The set of optimal decisions from D according to the criterion u_i (which will be denoted by D_i) is a fuzzy subset of D, since u_i is a fuzzy criterion. The membership function of D_i denoted by $\mu^{(i)}_{\max} : D \to [0,1]$ is defined, in accordance with Section 2.2, by

$$\mu^{(i)}_{\max}(d) = \inf\left(\nu_{\geq}(u_i(\rho(d)), u_i(\rho(d')) : d' \in D\right) \tag{12.10}$$

$$= \inf_{d' \in D} \left(\sup_{\substack{x, y \in R \\ x \geq y}} [\min(\mu_{i,d}(x), \mu_{i,d'}(y))] \right)$$

where $\nu_{\geq}$ is the membership function of the fuzzy relation "not less than" given by

$$\nu_{\geq}(a, b) = \sup\left[\min(\mu_a(x), \mu_b(y)) : x, y \in R, x \geq y\right] \tag{12.11}$$

for $a, b \in \mathbb{R}$.

Fuzzy sets D_i, $i = 1, \ldots, n$ were constructed independently on the ordering of criteria by their importance, which is typical for the lexicographic method. Now we define, with regard to the ordering, the fuzzy subsets $D^{(1)}, D^{(2)}, \ldots, D^{(n)}$ of D with membership functions $\pi_1, \pi_2, \ldots,$ $\pi_n : D \to [0,1]$ in the following inductive method. We say that

$$D^{(1)} = D_1, \quad \text{i.e.,} \quad \pi_1(d) = \mu^{(1)}_{\max}(d) \quad \text{for all } d \in D \tag{12.12}$$

$$D^{(i)} = D_i \cap D^{(i-1)}, \text{ i.e.,}$$

$$\pi_i(d) = \min\left(\mu_{\max}^{(i)}(d),\ \pi_{i-1}(d)\right), \qquad (12.13)$$
$$d \in D$$

for $i = 2, \ldots, n$. Then in the fuzzy set theoretical sense $D^{(i)} \supset D^{(i+1)}$, $i = 1, \ldots, n-1$. The set $D^{(n)}$ is the fuzzy set of lexicographically optimal decisions.

Up to now we have assumed that the ordering of criteria according to their importance was fixed. But even this ordering may be vague, expressed by a fuzzy ordering relation over the index set $\{1, \ldots, n\}$.

Let us suppose that this relation is represented by membership functions

$$\lambda_i : \{1, \ldots, n\} \to [0, 1], \quad i = 1, 2, \ldots, n \qquad (12.14)$$

where $\lambda_i(j)$ is the possibility that criterion u_i is at the jth position of importance. Then the set $D^{(1)}$ is a fuzzy subset of D with membership function

$$\pi_1(d) = \max_{i=1,\ldots,n}\left[\min(\mu_{\max}^{(i)}(d),\ \lambda_i(1)\right], \quad d \in D \qquad (12.15)$$

with an obvious interpretation (see Section 1.4). Evidently, if $\lambda_1(1)$, $\lambda_j(1) = 0, j = 2, \ldots, n$, then (12.15) turns into (12.12). Analogously, $D^{(j)}$, $j = 2, \ldots, n$, is a fuzzy subset of $D^{(j-1)}$ with membership function

$$\pi_j(d) = \min\left(\pi_{j-1}(d),\ \max_{i=1,\ldots,n}\left[\min(\mu_{\max}^{(i)}(d),\ \lambda_i(j))\right]\right) \qquad (12.16)$$

for any $d \in D$. Even in this case the degenerated fuzziness $\lambda_j(j) = 1$, $\lambda(k) = 0$, $j \neq k$, turns (12.16) into (12.13).

If only the ordering of criteria is fuzzy and expressed by the possibilities $\lambda_j(k)$ but the utility functions are crisp, then obviously $\mu_{\max}^{(i)}$ become $0-1$ characteristic functions and (12.15) and (12.16) turn into the special forms

$$\pi_1(d) = \max\left(\lambda_i(1) : i = 1, \ldots, n \text{ and } d \in D_i\right)$$

and

$$\pi_j(d) = \min_{k=1,\ldots,j}\left(\max(\lambda_i(k) : i = 1, \ldots, n \text{ and } d \in D_k)\right)$$

12.4 Fuzzy Pareto Optimum

Vector optimization methods for the vectors of fuzzy quantities from $\mathbb{R}^n$ can be generally based on some concepts discussed in Chapter 6, combined with the concepts of extremes from Section 2.2. These methods deal with vectors

$$u(\rho(d)) = (u_1(\rho(d)), \ldots, u_n(\rho(d))) \in \mathbb{R}^n, \quad d \in D$$

The domination concept (12.3) for fuzzy criteria changes into a fuzzy relation over D (or $\mathcal{C}$). Its membership function $\nu_{\succsim} : D \times D \to [0,1]$ is given by

$$\nu_{\succsim}(d, d') = \min \left[\min_{i=1,\ldots,n} (\nu_{\geq}(u_i(\rho(d)),\, u_i(\rho(d'))) \right. \tag{12.17}$$

$$\left. \max_{i=1,\ldots,n} (\nu_{>}(u_i(\rho(d))\,,\, u_i(\rho(d')))) \right], \quad d,\, d' \in D$$

where $\nu_{\geq} : R \times R \to [0,1]$, $\nu_{>} : R \times R \to [0,1]$ are membership functions of the ordering relations "not greater than" and "greater than" over real numbers, described in Section 2.2 by formulas (2.9) and (2.8).

Then the Pareto optimum P_D is a fuzzy subset of D with membership function $\pi_P : D \to [0,1]$, defined analogously to (12.4) and (12.5) by

$$\pi_P(d) = \inf_{d' \in D} \left(1 - \nu_{\succsim}(d', d)\right) \tag{12.18}$$

If $\nu_{>}(a,b) = 1 - \nu_{\geq}(b,a)$, $a,\, b \in R$ (which equality is not guaranteed, as shown by Example 1 in Section 2.2), then also

$$\pi_P(d) = 1 - \sup\left(\nu_{\succsim}(d', d) : d' \in D\right)$$

$$= 1 - \sup_{d' \in D} \left(\min \left[\min_{i=1,\ldots,n} [\nu_{\geq}(u_i(\rho(d')),\, u_i(\rho(d)))] \right.\right.$$

$$\left.\left. \max_{i=1,\ldots,n} [\nu_{>}(u_i(\rho(d')),\, u_i(\rho(d)))] \right] \right)$$

$$= 1 - \sup_{d' \in D} \left(\min \left[1 - \max_{i=1,\ldots,n} [\nu_{>}(u_i(\rho(d)),\, u_i(\rho(d')))] \right.\right.$$

$$\left.\left. \max_{i=1,\ldots,n} [\nu_{>}(u_i(\rho(d')),\, u_i(\rho(d)))] \right] \right)$$

Even the alternative Pareto optimum P_D^* based on the strong domination $\succ$ (see Section 11.2) can be modified for the fuzzy utilities. In fact, some discrepancies of P_D^* existing in the deterministic decision-making models are not critical in the fuzzy case, and the simplicity of the applied formulas is an evident advantage of such approach. Even in this case the strong domination $\succ$ turns into a fuzzy relation over D with membership function $\nu_{\succ} : D \times D \to [0,1]$, such that

$$\nu_{\succ}(d, d') = \min_{i=1,\ldots,n} (\mu_{>}(u_i(\rho(d)),\, u_i(\rho(d')))), \quad d,\, d' \in D \tag{12.19}$$

where $\mu_{>} : R \times R \to [0,1]$ is the membership function of the fuzzy relation "greater than" over real numbers. Then the set P_D^* is also a fuzzy subset of D with membership function $\pi^*_P : D \to [0,1]$,

$$\pi^*_P(d) = \inf_{d' \in D} (1 - \nu_{\succ}(d', d)) = 1 - \sup(\nu_{\succ}(d', d) : d' \in D)$$

12.5 Fuzzy Global Criterion

The theoretical tools developed for arithmetic processing of fuzzy quantities offer the best possibilities for fuzzification of equality (12.7), even if the representation theorem (see Section 2.3) may be applied also to more general forms of functions (12.6).

The arithmetics of fuzzy quantities can be applied to three forms of fuzzification of Equation 12.7. The first is based on the assumption used in the previous sections, namely that the utility functions u_i, $i = 1, \dots, n$ are fuzzy functions (which means that for any $d \in D$ the values $u_i(\rho(d))$ are fuzzy quantities) and that the weights $w_i \in R$, $i = 1, \dots, n$, are crisp numbers. Then (12.7) is a composition of crisp products and fuzzy additions

$$u(\rho(d)) = (w_1 \cdot u_1(\rho(d))) \oplus \cdots \oplus (w_n \cdot u_n(\rho(d))), \quad d \in D \tag{12.20}$$

The second form of fuzzification is the complementary one—we may assume that the weights of criteria are fuzzy quantities $w_i \in \mathbb{R}$, $i = 1, \dots, n$, and the utility functions are deterministic. Then the transcription of (12.7) is formally identical to (12.20), with transverted interpretation of fuzziness of the input quantities. It is, of course, acceptable to consider both the utilities $u_i(\rho(d))$ and the fuzzy weights w_i, $i = 1, \dots, n$, $d \in D$ to be fuzzy quantities. Then (12.7) becomes

$$u(\rho(d)) = [w_1 \odot u_1(\rho(d))] \oplus \cdots \oplus [w_n \odot u_n(\rho(d))], \quad d \in D \tag{12.21}$$

The assumptions concerning weights mean that for any w_i, its membership function $\mu_{w_i} : [0,1] \to [0,1]$, which means that $\mu_{w_i}(x) = 0$ for $x < 0$ or $x > 1$. All three types of fuzzification can be managed using the theoretical results of Chapters 3 and 5 if the natural ordering of operations (stressed by the brackets in Equations 12.20 and 12.21) is preserved to avoid misunderstandings connected with the distributivity rules. The maximization of the values of the fuzzy quantities $u(\rho(d)) \in \mathbb{R}$, $d \in D$ represents another version of the problem mentioned in previous sections; it is generally described in Section 2.2.

12.6 Dealing With a Fuzzy Manager

The general situations of multicriteria decision-making admit various specific modifications. One of them, introduced in [71], can be presented here as an example of such a modified decision-making problem with an interesting interpretation.

Let us consider a multicriteria decision-making problem, described as the global criterion optimization, where the global criterion $u : D \to R$ is defined as a convex combination

$$u(d) = w_1 \cdot u_1(d) + \cdots + w_n \cdot u_n(d), \quad d \in D \tag{12.22}$$

(we use the conventions in Equation (12.9)) and write $u_i(d)$, $u(d)$ instead of $u_i(\rho(d))$, $u(\rho(d))$, $i = 1, \ldots, n$). The decision-maker wishing to maximize the global utility $u(d)$ knows the deterministic particular utility functions $u_i : D \to R$ but does not know their weights.

The decision maker's only relevant information for the estimation of weights is vague comparative information about the success of particular, already accepted, decisions. This means — if the decision maker accepts decisions $d \in D$ and $d' \in D$, then vague information results if d was more successful than d' (i.e., $u(d) > u(d')$). The vagueness of the information is expressed by the possibility under which the inequality holds. The decision-maker's information sounds like this: "Decision d was not less successful than d' with possibility $\nu(d, d')$", where $\nu : D \times D \to [0, 1]$ is the nearer to 1.0 the higher the possibility is.

In this description the relation "not to be less successful than" is a fuzzy relation over D. It means that the estimation of the "true" weights $w_1, \ldots, w_n$ will be presented in fuzzy set theoretical concepts. To be the "true" vector of weights is a fuzzy property and the "true" vectors $(w_1, \ldots, w_n)$ form a fuzzy subset of the set

$$\mathcal{W} = \{\mathbf{w} = (w_1, \ldots, w_n) \in R^n : w_1 + \cdots + w_n = 1, w_i \geq 0, i = 1, \ldots, n\} \tag{12.23}$$

Using the terminology and notation of Section 6.1, the "true" weights vector $\mathbf{w}$ is a fuzzy vector quantity from $\mathbb{R}_n$.

Of course, the problem formulated above deals with a few dummy but essential assumptions.

First, we assume that the decision-maker is allowed to review the decision-making situation several times, theoretically without limitations. In each repetition an arbitrary decision can be accepted.

Second, there exists some subject evaluating the success of the already accepted decisions and informing the decision-maker (at least vaguely) about the comparative success. This subject may know the weights. We call this subject a *manager* even if this name can be misleading — the "manager" may also be impersonal, e.g., some natural or social process affected by the accepted decisions and giving a sequence of comparable results.

Third, we assume that the decision-maker is rational enough not to accept "bad" decisions, given the ability to recognize them. In our terms — since the decision-maker knows the values of utility functions u_i, $i = 1, \dots, n$, he is able to construct the Pareto optimum P_D (see Equations 12.4 and 12.5) and to choose the Pareto optimal decisions. Accepting this assumption we immediately assume that if the decision-maker chooses decisions d, $d' \in D$ and if $u_i(d) > u_i(d')$ for some $i \in \{1, \dots, n\}$, then there exists $j \in \{1, \dots, n\}$ such that $u_j(d') > u_j(d)$. The rationality of the decision-maker also means that, wishing to estimate the weights effectively, he does not choose decisions d, $d' \in D$ such that $u_i(d) = u(d')$ for all $i = 1, \dots, n$.

Let us suppose, now, that decisions d, $d' \in D$ were chosen and that, e.g., $u_i(d) > u_i(d')$ for some $i \in \{1, \dots, n\}$. Let us denote for a vector

$$\mathbf{w} = (w_1, \dots, w_n) \in \mathcal{W}$$

the number

$$r(i, \mathbf{w}, d, d') = \frac{\sum_{j \neq i,\, j=1}^{n} w_j (u_j(d') - u_j(d))}{u_i(d) - u_i(d')} \tag{12.24}$$

Then, according to (12.22), the manager's preference $u(d) \geq u(d')$ implies

$$1 \geq w_i \geq r(i, \mathbf{w}, d, d') \tag{12.25}$$

and on the other hand, the preference $u(d) < d(d')$ implies

$$0 \leq w_i < r(i, \mathbf{w}, d, d') \tag{12.26}$$

If $\nu(d, d')$ is the possibility of $u(d) \geq u(d')$, then (12.25) is valid with the possibility $\nu(d, d')$, and the possibility of (12.26) is $1 - \nu(d, d')$. We saw in Section 2.2 and recall from previous paragraphs that the relation $\nu_>(a, b) = 1 - \nu_\geq(b, a)$ does not generally hold. This fact concerns the fuzzy relations of ordering between fuzzy quantities a, $b \in \mathbb{R}$, where the possibilities $\nu_>(a, b)$ and $\nu_\geq(b, a)$ were computed from the membership functions μ_a and μ_b. In this section we consider the fuzzy ordering relation on D, where $\nu_>(d, d')$ is not computed but given by the manager. In this case the complementarity $\nu_\geq(d', d) = 1 - \nu_>(d, d')$ immediately follows from (1.1), and it is logically consistent and adequate for the modeled situation with respect to it. The "true" vector of weights is a fuzzy quantity w with a membership function $\mu_w : R^n \to [0, 1]$ (in fact $\mu_w : \mathcal{W} \to [0, 1]$ as $\mu_w(\mathbf{x}) = 0$ for $\mathbf{x} \in R^n - \mathcal{W}$). The membership function μ_w can be estimated by choosing a sequence of "experimental" decisions and by applying the following procedure.

If $d^{(1)}$, $d^{(2)} \in D$ were chosen, if $u_i(d^{(1)}) > u_i(d^{(2)})$ for some $i \in \{1, \ldots, n\}$ and if $u(d^{(1)}) \geq u(d^{(2)})$ with possibility $\nu(d^{(1)}, d^{(2)})$, then we construct a membership function $\mu^{(1,2)} : \mathcal{W} \to [0,1]$ using (12.25) and (12.26), and say that

$$\mu^{(1,2)}(x) = \nu\left(d^{(1)}, d^{(2)}\right)$$
$$\text{for } \mathbf{x} \in \mathcal{W} \cap \left\{\mathbf{y} \in R^n : y_i \geq r(i, \mathbf{y}, d^{(1)}, d^{(2)})\right\} \tag{12.27}$$
$$\mu^{(1,2)}(\mathbf{x}) = 1 - \nu\left(d^{(1)}, d^{(2)}\right)$$
$$\text{for } \mathbf{x} \in \mathcal{W} \cap \left\{\mathbf{y} \in R^n : y_i < r(i, \mathbf{y}, d^{(1)}, d^{(2)})\right\}$$

If the information about the possibility of both the inequalities $u(d^{(1)}) \geq u(d^{(2)})$ and $u(d^{(1)}) \leq u(d^{(2)})$ and their possibilities $\nu(d^{(1)}, d^{(2)})$ and $\nu(d^{(2)}, d^{(1)})$ is available, then the membership function $\mu^{(1,2)}$ achieves three values, namely

$$\mu^{(1,2)}(\mathbf{x}) = 1 - \nu\left(d^{(2)}, d^{(1)}\right)$$
$$\text{for } \mathbf{x} \in \mathcal{W} \cap \left\{\mathbf{y} \in R^n : y_i > r(i, \mathbf{y}, d^{(1)}, d^{(2)}\right\} \tag{12.28}$$
$$= 1 - \nu\left(d^{(1)}, d^{(2)}\right)$$
$$\text{for } \mathbf{x} \in \mathcal{W} \cap \left\{\mathbf{y} \in R^n : y_i < r(i, \mathbf{y}, d^{(1)}, d^{(2)}\right\}$$
$$= \nu\left(d^{(2)}, d^{(1)}\right) + \nu\left(d^{(2)}, d^{(1)}\right) - 1$$
$$\text{for } \mathbf{x} \in \mathcal{W} \cap \left\{\mathbf{y} \in R^n : y_i = r(i, \mathbf{y}, d^{(1)}, d^{(2)}\right\}$$

Having consequently chosen decisions $d^{(1)}$, $d^{(2)}, \ldots, d^{(m)}$, the decision-maker deals with several membership functions, such as those in (12.27) or (12.28). The "true" weights must fulfill the demands of all of them, which means that the fuzzy quantity w can be approximated by an intersection of the quantities (fuzzy subsets of $\mathcal{W}$) represented by membership functions $\mu^{(k,\ell)}$, k, $\ell = 1, \ldots, m$, $k \neq \ell$. Then the $(d^{(1)}, \ldots, d^{(m)})$-approximation of μ_w is a vector fuzzy quantity with membership function, denoted $\mu_w^{(m)} : \mathcal{W} \to [0,1]$, and defined by

$$\mu_w^{(m)}(\mathbf{x}) = \min\left(\mu^{(i,j)}(\mathbf{x}) : i = 1, \ldots, m-1, ; j = 2, \ldots, m, i < j\right),$$
$$\mathbf{x} \in \mathcal{W} \tag{12.29}$$

This function divides the set $\mathcal{W}$ into a finite number of disjoint convex and non-empty subsets, such that $\mu_w^{(m)}$ is constant over each of them.

Obviously $\mu^{(i,j)}(\mathbf{x}) = \mu^{(j,i)}(\mathbf{x})$ for all $\mathbf{x} \in \mathcal{W}$, which follows from (12.27) and (12.28).

The variability of the membership function $\mu_w^{(m)}$ also indicates the consistency of the manager's global preferences. Namely, the nearer the difference between the maximal and minimal values is to 1, the more determined the preferences are, and on the other hand, a small difference

$$\max\left(\mu_w^{(m)}(\mathbf{x}) : \mathbf{x} \in \mathcal{W}\right) - \min\left(\mu_w^{(m)}(\mathbf{x}) : x \in \mathcal{W}\right)$$

indicates the manager's hesitations on preferences. Moreover, if

$$\max\left(\mu_w^{(m)}(\mathbf{x}) : \mathbf{x} \in \mathcal{W}\right) < 1/2$$

then the manager's preferences are contradictory.

Example 41. Let us consider a 3-criteria decision-making procedure. Suppose that the decisions $d^{(1)}$, $d^{(2)}$, $d^{(3)}$ were chosen, where

$$\begin{aligned} u_1\left(d^{(1)}\right) &= 3, \quad u_2\left(d^{(1)}\right) = 3, \quad u_3\left(d^{(1)}\right) = 3 \\ u_1\left(d^{(2)}\right) &= 1, \quad u_2\left(d^{(2)}\right) = 2, \quad u_3\left(d^{(2)}\right) = 5 \\ u_1\left(d^{(3)}\right) &= 4, \quad u_2\left(d^{(3)}\right) = 2, \quad u_3\left(d^{(3)}\right) = 4 \end{aligned}$$

and let the information about the manager's preferences be as follows

$$\begin{aligned} u\left(d^{(1)}\right) \geq u\left(d^{(2)}\right) \quad &\text{with possibility} \quad \nu\left(d^{(1)}, d^{(2)}\right) = 0.85 \\ u\left(d^{(1)}\right) \leq u\left(d^{(2)}\right) \quad &\text{with possibility} \quad \nu\left(d^{(2)}, d^{(1)}\right) = 0.90 \\ u\left(d^{(3)}\right) \leq u\left(d^{(2)}\right) \quad &\text{with possibility} \quad \nu\left(d^{(2)}, d^{(3)}\right) = 0.80 \\ u\left(d^{(1)}\right) \geq u\left(d^{(3)}\right) \quad &\text{with possibility} \quad \nu\left(d^{(1)}, d^{(3)}\right) = 0.95 \end{aligned}$$

Then, using the relations presented above, we construct the approximate estimation of the vector of weights (w_1, w_2, w_3) as a 3-dimensional fuzzy quantity, which is a fuzzy subset of

$$\mathcal{W} = \left\{\mathbf{x} = (x_1, x_2, x_3) \in R^3 : x_1 + x_2 + x_3 = 1, \ x_i \geq 0, \ i = 1, 2, 3\right\}$$

with membership function $\mu_w^{(3)} : \mathcal{W} \to [0, 1]$. It is given by

$$\mu_w^{(3)}(\mathbf{x}) = \min\left(\mu^{(1,2)}(\mathbf{x}), \, \mu^{(2,3)}(\mathbf{x}), \, \mu^{(1,3)}(\mathbf{x})\right), \quad \mathbf{x} \in \mathcal{W}$$

where

$$\mu^{(1,2)}(\mathbf{x}) = 0.75 \quad \text{for } \mathbf{x} \in \mathcal{W},\ x_1 = 1/2 - 3x_3/4$$
$$= 0.15 \quad \text{for } \mathbf{x} \in \mathcal{W},\ x_1 > 1/2 - 3x_3/4$$
$$= 0.10 \quad \text{for } \mathbf{x} \in \mathcal{W},\ x_1 < 1/2 - 3x_3/4$$

$$\mu^{(2,3)}(\mathbf{x}) = 0.80 \quad \text{for } \mathbf{x} \in \mathcal{W},\ x_3 \geq 3 \cdot (1 - x_2)4$$
$$= 0.20 \quad \text{for } \mathbf{x} \in \mathcal{W},\ x_3 < 3 \cdot (1 - x_2)4$$

$$\mu^{(1,3)}(\mathbf{x}) = 0.95 \quad \text{for } \mathbf{x} \in \mathcal{W},\ x_2 \geq 1/2$$
$$= 0.05 \quad \text{for } \mathbf{x} \in \mathcal{W},\ x_2 < 1/2$$

Hence,

$$\mu_w^{(3)}(\mathbf{x}) = 0.75 \quad \text{for } \mathbf{x} = (1/8,\ 1/2,\ 3/8)$$
$$= 0.20 \quad \text{for } \mathbf{x} \in \mathcal{W},\ x_1 = 1/2 - 3x_2/4 \text{ and } x_2 \geq 1/2$$
$$= 0.15 \quad \text{for } \mathbf{x} \in \mathcal{W},\ x_1 > 1/2 - 3x_2/4 \text{ and } x_2 \geq 1/2$$
$$= 0.10 \quad \text{for } \mathbf{x} \in \mathcal{W},\ x_1 < 1/2 - 3x_2/4 \text{ and } x_2 \geq 1/2$$
$$= 0.05 \quad \text{for } \mathbf{x} \in \mathcal{W},\ x_2 < 1/2$$

If we map the set $\mathcal{W}$ as a triangle in R^3 with vertices (1,0,0), (0,1,0), (0,0,1), then it is divided into areas of constant values of $\mu_w^{(3)}$, as shown in Figure 12.2. Each area is completed by the corresponding value of the membership function $\mu_w^{(3)}$, and the general point (x_1, x_2, x_3) illustrates the system of coordinates in the triangular graph shown in Figure 12.2.

If the information about the manager's global preferences is extended and completed by the preference

$$u(d^{(2)}) \leq u(d^{(3)}) \quad \text{with possibility } \nu(d^{(3)},\ d^{(2)}) = 1.0$$

then functions $\mu^{(1,2)}$ and $\mu^{(1,3)}$ remain unchanged, and $\mu^{(2,3)} : \mathcal{W} \to [0, 1]$ is as follows

$$\mu^{(2,3)}(\mathbf{x}) = 0.80 \quad \text{for } \mathbf{x} \in \mathcal{W},\ x_3 = 3 \cdot (1 - x_2)/4$$
$$= 0.0 \quad \text{for } \mathbf{x} \in \mathcal{W},\ x_3 > 3 \cdot (1 - x_2)/4$$
$$= 0.20 \quad \text{for } \mathbf{x} \in \mathcal{W},\ x_3 < 3 \cdot (1 - x_2)/4$$

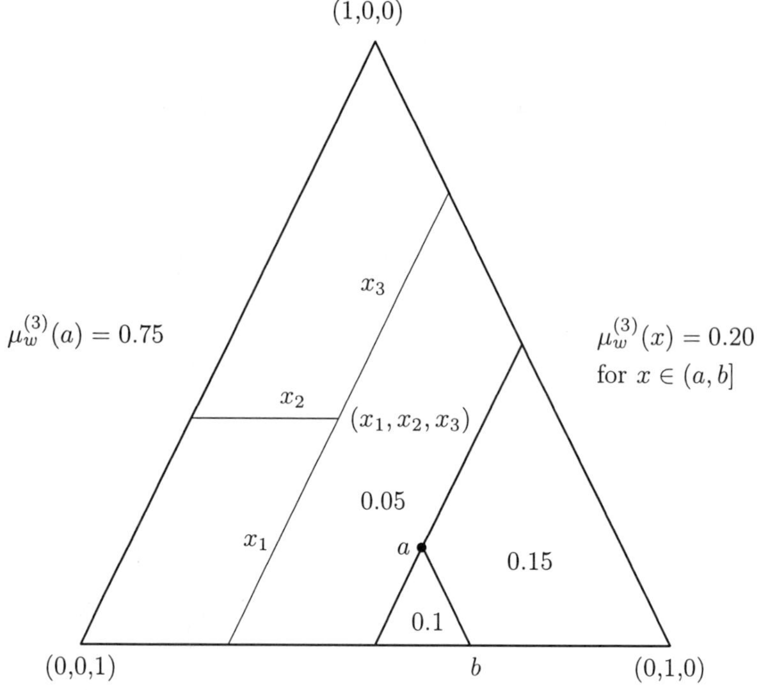

FIGURE 12.2
$a = (0.125,\ 0.5,\ 0.375), b = (0,\ 2/3,\ 1/3)$

and the $(d^{(1)}, d^{(2)}, d^{(3)})$-estimation $\mu_w^{(3)}$ of μ_w achieves values

$$
\begin{aligned}
\mu_w^{(3)} &= 0.75 \quad \text{for } \mathbf{x} = (1/8,\ 1/2,\ 3/8) \\
&= 0.15 \quad \text{for } \mathbf{x} \in \mathcal{W},\ x_3 \leq 3 \cdot (1 - x_2)/4,\ x_2 \geq 1/2, \\
&\qquad\quad\ \mathbf{x} \neq (0.125,\ 0.5,\ 0.375) \\
&= 0.05 \quad \text{for } \mathbf{x} \in \mathcal{W},\ x_3 \leq 3 \cdot (1 - x_2)/4,\ x_2 < 1/2 \\
&= 0.0 \quad\ \ \text{for } \mathbf{x} \in \mathcal{W},\ x_3 > 3 \cdot (1 - x_2)/4
\end{aligned}
$$

The corresponding subsets of $\mathcal{W}$ are shown in Figure 12.3.

The manager supposed in Example 41 evidently used a relatively consistent system of preferences. One point (weight vector (1/8, 1/2, 3/8)) in $\mathcal{W}$ has quite a high possibility, while the possibilities of the other weight vector are rather low. At any rate, none of the elements from $\mathcal{W}$ has a possibility equal to 1.

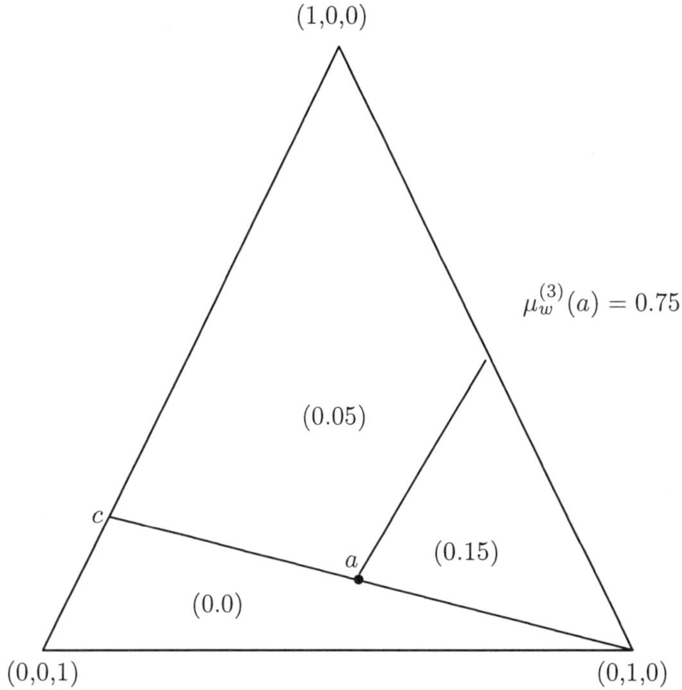

FIGURE 12.3
$a = (0.125,\ 0.5,\ 0.375), c = (0.25,\ 0,\ 0.75)$

It is evident from the foregoing discussion and example that even if some preferences are valid with possibility 1.0, the resulting fuzzy quantity w need not be regular, since its membership function μ_w does not fulfill condition (2.1). This is a consequence of the inconsistency of the manager's preferences present in the possibilities $\nu(d, d')$. Only if, for all pairs $d^{(i)}$, $d^{(j)}$ some of the values $\nu\left(d^{(i)},\ d^{(j)}\right)$ or $\nu\left(d^{(j)},\ d^{(i)}\right)$ were equal to 1.0, the approximation $\mu_w^{(m)}$ of μ_w, can the validity of (2.1) be guaranteed.

13

Fuzzy Cooperation

Fuzzy games of strategy became a widely investigated field of fuzzy set theory, and deserve a book unto themselves. Here we recall another application of the general concepts presented above to a small part of game theory, coalition games. Their fuzzy version can be described relatively briefly, and it recalls a specific branch of game theoretical thinking in practical applications frequently connected with vague phenomena.

13.1 Coalition Games

The general model of coalition games consists of the following components (see [73]). There is a non-empty and finite set of *players* I; each subset of $K \subset I$ of I is called a *coalition*. For every $K \subset I$, there is a set of achievable *pay-off vectors* $\mathcal{V}(K)$, characterizing the incomes of the members of coalition K. To simplify formal dealing with the game model, the sets $\mathcal{V}(K)$ are defined as cylindrical sets $\mathcal{V}(K) \subset R^I$ (where for every vector $\mathbf{y} = (y_i)_{i \in I} \in \mathcal{V}(K)$, only the coordinates y_i, $i \in K$, are significant and the others, y_j, $j \in I - K$, are arbitrary), fulfilling the following conditions:

$$\mathcal{V}(K) \text{ is closed subset of } R^I \tag{13.1}$$

$$\text{If } \mathbf{x} \in \mathcal{V}(K),\ \mathbf{y} \in R^I \text{ and } x_i \geq y_i \text{ for all } i \in K \text{ then } \mathbf{y} \in \mathcal{V}(K) \tag{13.2}$$

$$\mathcal{V}(K) \neq \emptyset \tag{13.3}$$

$$\mathcal{V}(K) = R^I \iff K = \emptyset \tag{13.4}$$

(the cylindricity of $\mathcal{V}(K)$ follows from Equation 13.2 immediately). The mapping $\mathcal{V}$ of the set of coalitions to the class of subsets of R^I is called the *characteristic function*.

To define the solution concept of such coalition games, the following notions are useful. If $\mathbf{x}$, $\mathbf{y} \in R^I$ and $K \subset I$, then we say that $\mathbf{x}$ *dominates*

$\mathbf{y}$ *via* K and write $\mathbf{x} \succsim_K \mathbf{y}$ iff

$$\begin{aligned} x_i &\geq y_i \quad \text{for all } i \in K, \text{ and} \\ x_y &> y_j \quad \text{for at least one } j \in K, \end{aligned} \tag{13.5}$$

(see the domination relation in Sections 12.2 and 12.3). Then it is possible to define for every $K \subset I$ the set $\mathcal{V}^*(K) \subset R^I$ such that

$$\begin{aligned} \mathcal{V}^*(K) &= \{\mathbf{y} \in R^I : \text{ there does not exist any } \mathbf{x} \in \mathcal{V}(K) \\ &\quad \text{such that } \mathbf{x} \succsim_K \mathbf{y}\} \\ &= \{\mathbf{y} = (y_i)_{i \in I} \in R^I : \text{ for every } \mathbf{x} \in \mathcal{V}(K) \text{ either} \\ &\quad \exists i \in K,\ y_i > x_i \text{or } \forall j \in K,\ y_j = x_j\} \end{aligned} \tag{13.6}$$

The intersection $\mathcal{V}(K) \cap \mathcal{V}^*(K) = P(K)$ is the *Pareto optimum* of the coalition K (see Sections 12.4 and 12.5).

The elementary and most natural solution concept in coalition game theory is called the *core* of the given game. It is a set of pay-off vectors $\mathcal{C} \subset R^I$ such that any $\mathbf{x} \in \mathcal{C}$ is a pay-off of the all-players coalition I, and it is not dominated via any coalition $K \subset I$ by any of its pay-off vectors. Symbolically,

$$\mathcal{C} = \mathcal{V}(I) \cap \left(\bigcap_{K \subset I} \mathcal{V}^*(K) \right) \tag{13.7}$$

A coalition game is called *superadditive* if, for any pair of disjoint coalitions $K,\ L \subset I,\ K \cap L = \emptyset$, the inclusion

$$\mathcal{V}(K \cup L) \supset \mathcal{V}(K) \cap \mathcal{V}(L) \tag{13.8}$$

holds (see [74]).

An important special class of coalition games is formed by so called games with *side-payments*, in which for every coalition $K \subset I$ there exists a real number $v_K \in R$ such that

$$\mathcal{V}(K) = \left\{ \mathbf{x} \in R^I : \sum_{i \in K} x_i \leq v_K \right\}$$

Then

$$\mathcal{V}^*(K) = \left\{ \mathbf{y} \in R^I : \sum_{i \in K} y_i \geq v_K \right\}, \quad P(K) = \left\{ \mathbf{x} \in R^I : \sum_{i \in K} x_i = v_K \right\}$$

and

$$\mathcal{C} = \left\{ \mathbf{x} \in R^I : \sum_{i \in I} x_i \leq v_I,\ \forall K \subset I \sum_{i \in K} x_i \geq v_K \right\}$$

The superadditivity in such games means that for $K,\ L \subset I,\ K \cap L = \emptyset$, the inequality $v_{K \cup L} \geq v_K + v_L$ holds.

The vagueness connected with coalition games is usually connected with the pay-off vectors and their sets $\mathcal{V}(K)$, and respectively with the values v_K in the side-payments case [13]. Such vague characteristic function can represent, e.g., a vague (fuzzy) distribution of common utility among players or vague (fuzzy) numerical utilities connected with some physical or material outcomes of the game. Their construction in many practical applications is connected with some subjective evaluation of the situation where the vague estimates are not rare.

13.2 Fuzzy Characteristic Function

The fuzzy pay-offs in general coalition games are represented by the fact that the sets $\mathcal{V}(K)$ are fuzzy subsets of R^I. In the terminology of Section 6.1, $\mathcal{V}(K)$ sets are similar to fuzzy vector quantities from $\mathbb{R}_n$. The difference from the concept given in Section 6.1 is caused by the contradiction between condition (6.2) and natural property (13.2) of the characteristic function. It can be verified that omitting condition (6.2) does not cause difficulties in this case, but to maintain the terminological purity, we speak about the fuzzy sets $\mathcal{V}(K) \subset_f R^I$ the same way we do about values of the fuzzy characteristic function $\mathcal{V}$. The membership function of $\mathcal{V}(K)$ is denoted by

$$\mu_{\mathcal{V}(K)} : R^I \to [0, 1]$$

To be fuzzy analogies of the coalition game theoretical characteristic function, the fuzzy sets $\mathcal{V}(K)$ have to fulfill some conditions analogous to (13.1), (13.2), (13.3), and (13.4) for any $K \subset I$. Condition (13.1) guarantees desirable topological properties of the border of the sets $\mathcal{V}(K)$, influencing, e.g., the non-emptiness of the Pareto optimum $P(K) = \mathcal{V}(K) \cap \mathcal{V}^*(K)$. For $\mathcal{V}(K) \subset_f R^I$, it takes the form

$$\text{The set } \left\{ \mathbf{x} \in R^I : \mu_{\mathcal{V}(K)}(\mathbf{x}) = 1 \right\} \text{ is closed and non-empty.} \tag{13.9}$$

The remaining three conditions are motivated by the game theoretical interpretation of $\mathcal{V}(K)$, and they are easily transformed into

$$\text{If } \mathbf{x},\ \mathbf{y} \in R^I,\ x_i \geq y_i \text{ for all } i \in K \text{ then } \mu_{\mathcal{V}(K)}(\mathbf{y}) \geq \mu_{\mathcal{V}(K)}(\mathbf{x}). \tag{13.10}$$

This condition means that for $\mathbf{x},\ \mathbf{y} \in R^I$, the equalities $x_i = y_i$ for all $i \in K$ imply $\mu_{\mathcal{V}(K)}(\mathbf{x}) = \mu_{\mathcal{V}(K)}(\mathbf{y})$, which guarantees the cylindricity of the sets $\mathcal{V}(K)$. The analogy of (13.3), which could be $\mu_{\mathcal{V}(K)}(\mathbf{x}) > 0$ for at least one $\mathbf{x} \in R^I$, follows from (13.9). The last analytic condition is

$$\begin{aligned} &\text{If } K \neq \emptyset \text{ then there exists } \mathbf{x} \in R^I,\ \mu_{\mathcal{V}(K)}(\mathbf{x}) = 0 \qquad (13.11)\\ &\text{if } K = \emptyset \text{ then } \mu_{\mathcal{V}(K)}(\mathbf{x}) = 1 \text{ for all } \mathbf{x} \in R^I \end{aligned}$$

The domination relation $\succsim_K$ via a coalition K does not turn into a fuzzy relation (as in Section 12.4), since the input vectors $\mathbf{x}$, $\mathbf{y}$ are not fuzzy quantities but crisp vectors, with a certain probability of belonging to some fuzzy sets. Nevertheless, the set $\mathcal{V}^*(K)$ defined by (13.6) turns into a fuzzy subset of R^I, since the membership of some $\mathbf{y} \in R^I$ to $\mathcal{V}^*(K)$ depends on its domination by some vectors, whose membership in the fuzzy set $\mathcal{V}(K)$ is not deterministic.

If $\mathbf{x}, \mathbf{y} \in R^I$, $K \subset I$, and $\mathbf{x} \succsim_K \mathbf{y}$, then the possibility that $\mathbf{y}$ is dominated via K by $\mathbf{x}$ from $\mathcal{V}(K)$ is equal to the possibility $\mu_{\mathcal{V}(K)}(\mathbf{x})$ of $\mathbf{x}$ being an element of $\mathcal{V}(K)$. The possibility that $\mathbf{y} \in R^I$ is dominated by some vector from $\mathcal{V}(K)$ is then equal to

$$\sup\left(\mu_{\mathcal{V}(K)}(\mathbf{x}) : \mathbf{x} \succsim_K \mathbf{y}\right)$$

and, consequently, the membership function $\mu^*_{\mathcal{V}(K)} : R^I \to [0,1]$ of the fuzzy set $\mathcal{V}^*(K) \subset_f R^I$ is given by

$$\begin{aligned} \mu^*_{\mathcal{V}(K)}(\mathbf{y}) &= 1 - \sup\left(\mu_{\mathcal{V}(K)}(\mathbf{x}) : \mathbf{x} \succsim_K \mathbf{y}\right) \\ &= \inf\left(1 - \mu_{\mathcal{V}(K)}(\mathbf{x}) : \mathbf{x} \succsim_K \mathbf{y}\right) \end{aligned} \qquad (13.12)$$

Condition (13.10) and equality (13.12) imply that if $\mathbf{x}, \mathbf{y} \in R^I$, $x_i \geq y_i$ for all $i \in K$, then

$$\mu^*_{\mathcal{V}(K)}(\mathbf{x}) \geq \mu^*_{\mathcal{V}(K)}(\mathbf{y})$$

The Pareto optimum $P(K) = \mathcal{V}(K) \cap \mathcal{V}^*(K)$ of the coalition $K \subset I$ is then a fuzzy subset of R^I with the membership function $\mu_{P(K)} : R^I \to [0,1]$, defined by

$$\begin{aligned} \mu_{P(K)}(\mathbf{x}) &= \min\Big(\mu_{\mathcal{V}(K)}(\mathbf{x}),\ \mu^*_{\mathcal{V}(K)}(\mathbf{x})\Big) \qquad (13.13)\\ &= \min\left(\mu_{\mathcal{V}(K)}(\mathbf{x}), \left(1 - \sup(\mu_{\mathcal{V}(K)}(\mathbf{y}) : \mathbf{y} \succsim_K \mathbf{x})\right)\right) \end{aligned}$$

for any $\mathbf{x} \in R^I$.

13.3 Fuzzy Core

It is evident also that the core $\mathcal{C}$ of the fuzzy coalition game under consideration (in the deterministic case defined by Equation 13.7) is a fuzzy subset of R^I. Its membership function is denoted by $\mu_C : R^I \to [0,1]$ and defined, analogously to (13.7), by

$$\mu_C(\mathbf{x}) = \min_{K \subset I} \left(\mu_{\mathcal{V}(I)}(\mathbf{x}),\ \mu^*_{\mathcal{V}(K)}(\mathbf{x}) \right) \tag{13.14}$$
$$= \min_{K \subset I} \left(\mu_{\mathcal{V}(I)}(\mathbf{x}),\ \left(1 - \sup(\mu_{\mathcal{V}(K)}(\mathbf{y}) : \mathbf{y} \succsim_K \mathbf{x})\right)\right)$$

The superadditivity formula (13.8), which is very significant for practical properties and the existence of the core in deterministic coalition games, achieves in the fuzzy case the following form. If $K, L \subset I,\ K \cap L = \emptyset$, then

$$\mu_{\mathcal{V}(K \cup L)}(\mathbf{x}) \geq \min \left(\mu_{\mathcal{V}(K)}(\mathbf{x}),\ \mu_{\mathcal{V}(L)}(\mathbf{x}) \right) \tag{13.15}$$

for all $\mathbf{x} \in R$. It means, by the way, that for any $K \subset I$, and any $\mathbf{x} \in R^I$,

$$\mu_{\mathcal{V}(I)}(\mathbf{x}) \geq \min \left(\mu_{\mathcal{V}(K)}(\mathbf{x}),\ \mu_{\mathcal{V}(I-K)}(\mathbf{x}) \right)$$

In deterministic coalition game theory, the core can be empty. Theoretically, it might happen even in the fuzzy coalition game model, and then $\mu_C(\mathbf{x}) = 0$ for all $\mathbf{x} \in R^I$. Analogously, the core may be such that $\mu_C(\mathbf{x}) = 1$ for some $\mathbf{x} \in R^I$, and in such a case the existence of the core is guaranteed. But the most common case that cannot appear in the deterministic model is that there exists a non-empty set

$$\left\{ \mathbf{x} \in R^I : 0 < \mu_C(\mathbf{x}) < 1 \right\}$$

of the pay-off vectors, which may be the possible core-like solution of the given game; the probability depends on the fuzzy properties of the sets $\mathcal{V}(K)$.

13.4 Fuzzy Side-Payments

Let us consider the side-payments coalition games, mentioned in Section 13.1, where the numbers v_K, $K \subset I$, are fuzzy quantities from $\mathbb{R}$. Our problem now is to find the structure of the set $\mathcal{V}(K)$ and of the other sets derived from $\mathcal{V}(K)$, including the core. If we denote by $\pi_K : R \to [0,1]$ the membership function of the fuzzy quantity $v_K \in \mathbb{R}$, then the membership function $\mu_{\mathcal{V}(K)}$ of the fuzzy set $\mathcal{V}(K) \subset_f R^I$ is derived from π_K in the following way. The possibility that $\mathbf{x} \in \mathcal{V}(K)$ is equal to the possibility that v_K achieves a value $y \in R$, such that $y \geq \sum_K x_i$, which is the possibility

$$\sup \left(\pi_K(z) : z \in R,\ z \geq y \right)$$

This means that

$$\mu_{\mathcal{V}(K)}(\mathbf{x}) = \sup\left(\pi_K(z) : z \in R, \sum_{i\in K} x_i \leq z\right) \tag{13.16}$$

This formula guarantees the validity of conditions (13.9), (13.10), and (13.11) for $\mathcal{V}(K)$ and, moreover, it preserves its linearity, since for any $r \in (0,1)$ there exists $z \in R$ such that

$$\{\mathbf{x} \in R^I : \mu_{\mathcal{V}(K)}(\mathbf{x}) \geq r\} = \left\{\mathbf{x} \in R^I : \sum_{i\in K} x_i \geq z\right\}$$

The Pareto optimum is represented by the intersection (13.13), and it preserves the linearity property in a certain sense. The core, defined by (13.14) with respect to (13.16), is characterized by the membership function

$$\begin{aligned}
\mu_C(\mathbf{x}) &= \min_{K\subset I}\left(\mu_{\mathcal{V}(I)}(\mathbf{x}),\, \mu^*_{\mathcal{V}(K)}(\mathbf{x})\right) \\
&= \min_{K\subset I}\left(\sup\left[\pi_I(z) : z \in R, \sum_{i\in I} x_i \leq z\right],\right. \\
&\quad \left.\sup\left[\pi_K(z) : z \in R, \sum_{i\in K} x_i \geq z\right]\right)
\end{aligned}$$

since $\mu_{\mathcal{V}(K)}(\mathbf{x}) = \sup\left(\pi_K(z) : z \in R, \sum_{i\in K} x_i \geq z\right)$.

14

The Structure of Fuzziness — Heuristic Views

Some properties of fuzzy quantities, such as their normality (2.1), symmetry (or transversibility), and additive or multiplicative equivalence represent not only technical tools suitable for handling vague data, but also offer some perception of the inner structure of fuzziness and vagueness, at least if numerical phenomena are considered.

14.1 Meaning of Normality

The normality assumption (2.1) placed on fuzzy quantities, demanding that

$$\sup\left(\mu_a(x) : x \in R\right) = 1$$

for $a \in \mathbb{R}$, is often used in a more common context than that of numerical fuzzy quantities. Some authors demand for every fuzzy set A the existence of some non-empty core of elements $x \in \mathcal{U}$ fulfilling $\mu_a(x) = 1$. In the case of fuzzy quantities with $\mathcal{U} = R$, this means the assumption of the existence of such numerical values $x \in R$ for any $a \in \mathbb{R}$ that are certainly values of a. Such values are also called the *modal values* of a. The fuzzy set is composed of this crisp core surrounded by some additional vague appendices, and the same is true for fuzzy quantities. In this book we accept this approach (motivated here by formal reasons), as shown in (2.1).

At any rate, the positive value of

$$\sup\left(\mu_a(x) : x \in R\right)$$

or more generally

$$\sup(\mu_A(x) : x \in \mathcal{U})$$

where $a \in \mathbb{R}$ or $A \subset_f \mathcal{U}$, using the notation of Section 1.2, determines the possibility that the quantity a achieves at least some real value, or that the vague property modeled by the fuzzy set A appears for at least one element of the universum. Limiting our further considerations to the fuzzy quantities or to vague numerical values, the supremum in (2.1) evaluates the possibility that the fuzzy quantity discussed can really exist. Such existence is not always guaranteed (see the properties of game theoretical core in Section 13.3 or the manager's "true" weights of criteria in Section 12.6) and in this sense the consequent validity of (2.1) need not be self-evident. From the formal point of view, condition (2.1) is necessary, namely for correct understanding of the equivalences $\sim_\oplus$ and $\sim_\odot$. If the condition is not fulfilled, then very dissimilar fuzzy quantities can be found to be equivalent if 0-symmetric fuzzy quantities with sufficiently small values of $\sup(\mu_s(x) : x \in R)$ are used in the definition of $\sim_\oplus$ (and analogously 1-transversible quantities in the definition of $\sim_\odot$). This fact is reflected even in [64] and [66], where general fuzzy quantities are normalized before the principles of (4.4) are applied in the definition of additive equivalence. In other parts of the theoretical and practical considerations presented in the foregoing chapters, assumption (2.1) is not unavoidable, and the interpretation of the supremum discussed here becomes more significant.

14.2 Additive Decomposition

As mentioned above, the supremum value of the membership function expresses the uncertainty under which the fuzzy quantity being discussed even exists in the investigated model of some vague situation. This uncertainty is not the only one incorporated in the membership functions of fuzzy quantities.

Some fuzzy quantities can be additively decomposed. If $a \in \mathbb{R}$, then there exists $a^{(1)} \in \mathbb{R}$ and $s^{(1)} \in \mathbb{S}_0$ such that $a = a^{(1)} \oplus s^{(1)}$. The same procedure may be repeated, and a sequence of pairs $\left(a^{(i)}, s^{(i)}\right)$, $i = 1, 2, \ldots$, may be constructed such that

$$a^{(i)} = a^{(i+1)} \oplus s^{(i+1)}, \; s^{(i+1)} \in \mathbb{S}_0, \; i = 1, 2, \ldots .$$

Each addition operation in the described procedure increases the support set of $a^{(i)}$ in comparison with the support set of $a^{(i+1)}$ (unless $s^{(i+1)} = \langle 0 \rangle$), as shown in [66]. This means that at least for fuzzy quantities with finite support, the procedure leads after a finite number of steps to a terminal decomposition

$$a = a^{(0)} \oplus s^{(0)}, \quad s^{(0)} \in \mathbb{S}_0$$

such that $a^{(0)}$ cannot be already decomposed, except for the pseudo-decomposition $a^{(0)}=a^{(0)} \oplus \langle 0 \rangle$.

But even in the case of more general fuzzy quantities $a \in \mathbb{R}$, the decomposition

$$a = a^{(1)} \oplus s, \quad s \in \mathbb{S}_0 \tag{14.1}$$

is correct. In some rare cases the single s fulfilling (14.1) is $s = \langle 0 \rangle$, even in the case of a quite complicated "residual" fuzzy quantity $a^{(1)}$. On the other hand, if $a \in \mathbb{S}_y$ for some $y \in R$, then $a = \langle y \rangle \oplus s$, $s \in \mathbb{S}_0$, which represents the opposite extreme.

If we try to interpret decomposition (14.1), it is necessary to remember that 0-symmetric fuzzy quantities from $\mathbb{S}_0$ represent some kind of "fuzzy zeros". Their vagueness is symmetrically distributed and they represent in a certain sense a balanced symmetric component of fuzziness hidden in a. On the other hand, the remaining member of (14.1), the residual fuzzy quantity $a^{(1)}$ (especially if it is further indecomposable), represents the asymmetric, unbalanced, and in some sense chaotic and irregular component of vagueness, represented by a.

Summing up the previous considerations, we see that each fuzzy quantity $a \in \mathbb{R}$ contains three kinds of vagueness:

- The uncertainty of the mere existence of a in the real world, represented by $\sup(\mu_a(x) : x \in R)$.
- Symmetric regular vagueness, concerning the extent of possible values of a and the concentration or decentralization of the distribution of vagueness around some inner core of its membership function; this symmetric vagueness is represented by the symmetric component $s \in \mathbb{S}_0$ of the decomposition (14.1). Roughly speaking, the component $s \in \mathbb{S}_0$ displays "how fuzzy the fuzzy zero is" contributing to the definitive form of a in (14.1).
- The "core" of vagueness; irregular asymmetric uncertainty concerning the structure of chaotic distribution of possibilities, represented by the asymmetric component $a^{(1)}$ of the decomposition (14.1).

It was shown in the previous parts of this book, especially in Chapters 4 and 5 (and in Section 14.1), that each of these kinds of vagueness has its specific properties and obeys specific rules. This shows that they represent rather different or at least differently viewed aspects of the common notion of vagueness.

14.3 Multiplicative Decomposition

It is worth mentioning that, analogously to the additive decomposition of fuzzy quantities from $\mathbb{R}$, the multiplicative decomposition should be considered, at least for signed fuzzy quantities from $\mathbb{R}^*$. If $a \in \mathbb{R}^*$, then there generally exist $a^{(1)} \in \mathbb{R}^*$ and $t \in \mathbb{T}_1$ such that

$$a = a^{(1)} \odot t \tag{14.2}$$

with the extreme cases represented by the possibility that either the single t fulfilling (14.2) is $t = \langle 1 \rangle$ or, on the other hand, that $a^{(1)} = \langle y \rangle$ for some $y \neq 0$ if $a \in \mathbb{T}_y$. In this case the decomposition means that, in addition to the supremum of the membership function, the interpretation of which we already know, there exist two types of vagueness that can be derived from $a \in \mathbb{R}^*$. The (in some sense) regular or balanced one is represented by the transversible component of (14.2), showing "how fuzzy the fuzzy unit is," presented in the decomposed fuzzy quantity; irregular vagueness is represented by $a^{(1)}$, which reflects the proper uncertainty hidden in a.

The existence of two possible decompositions of given fuzzy quantities (and some other decompositions cannot be excluded) provokes some questions. Namely, which is the "main" one? The answer cannot be absolute. Each has meaning in the context of the model of a given vague quantity, such as fuzzy contaminated numeric data or other fuzzy quantitative phenomena. The application and concrete interpretation of the fuzzy quantity being considered justifies the eventual choice of the concrete type of decomposition.

Much more significant is the fact that each of the suggested decompositions (14.1) and (14.2) separates two components from a general fuzzy quantity. One is regular, represents the extent of the uncertainty, and has its balanced concentration or dispersion around some type of an important center (zero or unit). The second is irregular, may be chaotic, and reflects the unbalanced inner uncertainty incorporated in the fuzzy quantity being considered. It is quite attractive to consider this second component for the essential representation of fuzziness. This attitude can be supported by the philosophy of Chapters 4 and 5, which consider the 0-symmetric, respectively 1-transversible, fuzzy quantities for some kind of negligible "noise" associated with the really interesting arithmetically processed fuzzy quantities.

List of Main Symbols

Symbol	Definition
$\mathcal{U}$	universum
χ_A	characteristic function
μ_A	membership function
$A \subset_f \mathcal{U}$	fuzzy subset
$\overline{A}$	complement
$A \cup B,\ A \cap B$	union, intersection
$\emptyset$	empty set
$\neg s$	negation of statement
$s \wedge t$	conjunction of statements
$s \vee t$	disjunction of statements
R	set of real numbers
$R_0 = R - \{\emptyset\}$	set of non-zero real numbers
μ_a	membership function
$a,\ b,\ c, \ldots$	fuzzy quantities
$\mathbb{R}$	set of fuzzy quantities
$-a$	opposite fuzzy quantity
$1/a$	reciprocal zero quantity
$\mathbb{R}_0$	set of non-zero fuzzy quantities
$\langle x \rangle$	degenerated fuzzy quantity
$\succ,\ \succsim,\ \sim$	ordering relations
$a \oplus b$	sum of fuzzy quantities
$a \odot b$	product of fuzzy quantities
$\mathbb{R}^+$	set of positive fuzzy quantities
$\mathbb{R}^-$	set of negative fuzzy quantities
$\mathbb{R}^*$	set of signed fuzzy quantities
$r \cdot a$	crisp product
a^r	root with crisp exponent
a^b	root with fuzzy exponent
$\mathbb{S}_y$	set of y-symmetric fuzzy quantities
$\mathbb{S}$	set of symmetric fuzzy quantities
$\sim_\oplus$	additive equivalence
$\mathbb{T}_y$	set of y-transversible fuzzy quantities

Symbol	Definition
$\mathbb{T}$	set of transversible fuzzy quantities
$\sim_{\odot}$	multiplicative equivalence
$\approx$	weak equivalence
R^n	set of n-dimensional real vectors
$\mathbf{x}$, $\mathbf{y}$	real-valued vector
$\mathbb{R}_n$	set of n-dimensional fuzzy vector quantities
$\mathbf{T}$	real-valued matrix
$\mathbb{R}^n$	set of n-dimensional vectors of fuzzy quantities
$\mathbf{a}$, $\mathbf{b}$	vector of fuzzy quantities
$\mathbb{R}^\wedge$	set of extended fuzzy quantities
$a \boxplus b$	convolutive sum of fuzzy quantities
$a \boxdot b$	convolutive product of fuzzy quantities
$\mathcal{L}$	lattice
$\mathbb{R}_{\mathcal{L}}$	set of $\mathcal{L}$-fuzzy quantities
$m_{A,x,}$, $\mu_{A,x}$	ultra-fuzzy membership functions
ρ	resulting function
D	set of decisions
C	set of consequences
u_i, u	utility functions
P_C, P_D	Pareto optimum
$\mathcal{V}(K)$	set of pay-offs
$\mathcal{C}$	core

Literature

Selected Journals

Automatica, Journal of IFAC. Pergamon Press, Oxford–Elmsford, New York.

BUSEFAL. Université Paul Sabatier, Toulouse.

European Journal of Operational Research. North–Holland, Amsterdam.

Fuzzy Sets and Systems. North–Holland, Elsevier, Amsterdam.

Fuzzy Systems and A. I. Magazine. Romanian Society for Fuzzy Systems. Iasi University Press, Iasi, Romania.

IEEE Transactions on Fuzzy Systems. IEEE, Piscataway, New Jersey.

IEEE Transactions on Systems, Man and Cybernetics. IEEE, New York.

Information and Control. Chinese Automation Society, Shenyang, China.

Information Sciences. Elsevier, New York.

International Journal of Approximative Reasoning. Elsevier, New York.

International Journal of Control. Taylor and Francis, London.

International Journal of Uncertainty, Fuzziness and Knowledge-Based Systems. World Scientific, London.

International Journal on General Systems. Gordon and Breach, Philadelphia.

Japanese Journal of Fuzzy Theory and Systems. SOFT, Tokyo. Allerton Press, New York.

Journal of Fuzzy Logic and Intelligent Systems. Korea Fuzzy Math. and Syst. Soc., Seoul.

The Journal of Fuzzy Mathematics. Int. Fuzzy Math. Inst., Los Angeles.

Journal of Intelligent and Fuzzy Systems: Applications in Engineering and Technology. J. Wiley, New York.

Journal of Mathematical Analysis and Applications. Academic Press, Duluth, Minnesota.

Kybernetes. MCB University Press, Bradford, Great Britain.

Kybernetika. Academia, Prague.

Bibliography

[1] **Albrycht, J. and Wisniewski, H. (Eds.)**, *Proceedings of the Polish Symposium on Interval and Fuzzy Mathematics*, August 1983, Institute of Mathematics, Technical University of Poznañ, Poznañ, 1985.

[2] **Badard, R.**, Fixed point theorems for fuzzy members, *Fuzzy Sets Systems* 13, 291–302, 1984.

[3] **Bellman, R. and Zadeh, L. A.**, Decision making in fuzzy environment, *Manag. Sci.*, 17, 141–164, 1970.

[4] **Bezdek, J. C.**, *Pattern Recognition With Fuzzy Objective Functions*, Plenum, New York, 1981.

[5] **Bezdek, J. C. (Ed.)**, *Analysis of Fuzzy Information, Vol. I — Mathematics and Logic*, CRC Press, Boca Raton, 1987.

[6] **Bezdek, J. C. (Ed.)**, *Analysis of Fuzzy Information, Vol. II — Artificial Intelligence and Decision Systems*, CRC Press, Boca Raton, 1987.

[7] **Bezdek, J. C. (Ed.)**, *Analysis of Fuzzy Information, Vol. III — Applications in Engineering and Sciences*, CRC Press, Boca Raton, 1987.

[8] **Billot, A.**, *Economic Theory of Fuzzy Equilibria, An Axiomatic Analysis.* Lecture Notes in Economics and Mathematical Systems 373, Springer-Verlag, Berlin, 1992.

[9] **Black, M.**, Vagueness: an exercise in local analysis, *Philosophy Science*, 4, 427–455, 1937.

[10] **Bocklisch, St., Orlovski, S., Peschel, M., and Nishiwaki, Y. (Eds.)**, *Fuzzy Set Applications, Methodological Approaches and Results*, Akademie–Verlag, Berlin, 1986.

[11] **Buckley, J. J.**, Fuzzy hierarchical analysis, *Fuzzy Sets Systems* 17, 233–247, 1985.

[12] **Butnariu, D.**, Fuzzy games. A description of the concept, *Fuzzy Sets Systems* 1, 181–192, 1978.

[13] **Butnariu, D.**, Values and cores of fuzzy games with infinitely many players, *Int. J. Game Theory* 16, 43–68, 1987.

[14] **Butnariu, D.**, Stability and Shapley value for n-persons fuzzy games, *Fuzzy Sets Systems* 7, 191–207, 1982.

[15] **Campos, L. and Gonzales, A.**, Fuzzy matrix games, considering the criteria of the players, *Kybernetes* 20 (1), 17–28, 1991.

[16] **Černý, M. and Glűckaufová, D.**, Fuzzy concepts in multiaspect decision making, *Ekonomicko-matematický obzor*, 23, 55–65, 1987.

[17] **Černý, M., Glűckaufová, G., and Loula, D. (Eds.)**, *Proceedings of the International Workshop: Multicriteria Decision Making, Methods — Algorithms — Applications, Liblice, March 18–22, 1991*, Inst. of Economics, Czechoslovak Academy of Sciences, Prague, 1992.

[18] **Chanas, S. and Kolodziejczyk, W.**, Maximum flow in a network with fuzzy arc capacities, *Fuzzy Sets Systems* 8, 165–179, 1982.

[19] **Delgado, M., Verdegay, J. L., and Vila, M. A.**, A general model for fuzzy linear programming, *Fuzzy Sets Systems* 29, 21–29, 1989.

[20] **Dijkman, J., van Haeringern, H., and de Lange, S.**, Fuzzy numbers, *Journal Math. Anal. Appl.*, 92, 301, 1983.

[21] **Dinola, A., and Ventre, A. G. S.**, *The Mathematics of Fuzzy Systems*, Verlag TUV, Rheinlan, 1986.

[22] **Dubois, D.**, Belief structure, possibility theory and decomposable confidence measures on finite sets, *Computers Artificial Intelligence* 5, 403–416, 1986.

[23] **Dubois, D. and Prade, H.**, Fuzzy numbers: An overview, in *Analysis of Fuzzy Information*, Vol. 2, Bezdek, J. C., Ed., CRC Press, Boca Raton, 1988, 3–39.

[24] **Dubois, D. and Prade, H.**, *Fuzzy Sets Systems*, Academic Press, New York, 1980.

[25] **Dubois, D. and Prade, H.**, *Possibility Theory*, Plenum, New York, 1988.

[26] **Evans, G. W., Karwowski, W., and Wilhelm, M. R. (Eds.)**, *Fuzzy Methodologies for Industrial and System Engineering*, Elsevier, Amsterdam, 1989.

[27] **Florenzo, M.**, Edgeworth equilibria, fuzzy core, and equilibria of a production economy without ordered preferences, *J. Math. Anal. Appl.* 153, 18–36, 1990.

[28] **Goguen, J. A.**, L-fuzzy sets, *J. Math. Anal. Appl.* 18, 145–174, 1967.

[29] **Gupta, M. M. (Ed.)**, *Fuzzy Automata and Decision Processes*, North–Holland, New York, 1977.

[30] **Gupta, M. M. and Sanchez, E. (Eds.)**, *Fuzzy Information and Decision Processes*, North–Holland, Amsterdam, 1982.

[31] **Gupta, M. M. and Sanchez, E. (Eds.)**, *Approximate Reasoning in Decision Analysis*, North–Holland, Amsterdam, 1982.

[32] **Gupta, M. M., Ragade, R. K., and Yager, R. S. (Eds.)**, *Advances in Fuzzy Set Theory and Applications*, North–Holland, Amsterdam, 1979.

[33] **Hájek, P., Havránek, T., and Jiroušek, R.**, *Uncertain Information Processing in Expert Systems*, CRC Press, Boca Raton.

[34] **Harman, B.**, Sum and product of the modified real fuzzy numbers, *Kybernetika*, supplement, 1992.

[35] **Harman, B.**, On associativity of the product of modified fuzzy numbers, *Tatra Mountains Math. Pub.* 1, 45–49, 1992.

[36] **Höhle, U.**, Representation theorem for L-fuzzy quantities, *Fuzzy Sets Systems* 3 (1), 83–107, 1981.

[37] **Jones, A., Kaufmann, A., and Zimmermann, H. J.**, *Fuzzy Set Theory and Applications*, D. Reidel, Dordrecht, 1985.

[38] **Kacprzyk, J.**, *Multistage Decision-Making Under Fuzziness, ISR Interdisciplinary Systems Research Vol. 79*, TUV Rheinland, Düsseldorf, 1983.

[39] **Kacprzyk, J. (Ed.)**, *Management Decision Support Systems Using Fuzzy Sets and Possibility Theory*, Springer-Verlag, Berlin, 1985.

[40] **Kacprzyk, J. and Fedrizzi, M., (Eds.)**, *Fuzzy Regression Analysis*, Omnitech Press, Physica-Verlag, Warsaw, 1992.

[41] **Kang In Ku (Ed.)**, *Proceedings of the Fifth IFSA'93 Congress, Seoul.* MFKMS, Chungang University, Seoul, 1993.

[42] **Kaufmann, A.**, Hybrid convolution, a way to combine fuzzy numbers and random variables, *Fuzzy Mathematics* (Huazhong, China), 1 (2), 1, 1981.

[43] **Kaufmann, A. and Gupta, M. M.**, *Introduction to Fuzzy Arithmetics*, Van Nostrand, New York, 1985.

[44] **Kelley, J. E.**, Critical-path planning and scheduling: mathematical basis, *Operational Research*, IX, 4, 1961.

[45] **Keresztfalvi, T. and Kovács, M.**, q, p-Fuzzification of arithmetic operations, *Tatra Mountain Math. Publ.*, 1, 65–71, 1992.

[46] **Keresztfalfi, T. and Rommelfanfger, H.**, Multicriteria fuzzy optimization based on Yager's parametrized t-norm, *Found. Comp. Decis. Making Syst.*, 16 (2), 99–110, 1991.

[47] **Klement, E. P. (Ed.)**, *Proceedings of the Third International Seminar on Fuzzy Set Theory at J. Kepler University (1981)*, Johannes Kepler Universität, Linz, 1981.

[48] **Klir, G. J.**, Where do we stand on measures of uncertainty, ambiguity, fuzziness and the like? *Fuzzy Sets Systems* 24, 141–160, 1987.

[49] **Klir, G. J. and Folger T. A.**, *Fuzzy Sets, Uncertainty, and Information*, Prentice Hall, Englewood Cliffs, 1988.

[50] **Knoptmacher, J.**, On measures of fuzziness, *J. Math. Anal. Appl.* 49, 529–534, 1957.

[51] **Kohout, L.**, *A Perspective on Intelligent Systems, Chapman & Hall*, London, 1990.

[52] **Kovács, M. and Tran, L. H.**, Algebraic structure of centered M-fuzzy numbers, *Fuzzy Sets Systems* 39 (1), 91–100, 1991.

[53] **Kramosil, I.**, From an alternative model of rough sets to fuzzy sets, *Kybernetika*, supplement, 1992.

[54] **Kruse, R. and Mayer, K. D.**, *Statistics with Vague Data*, D. Reidel, Kluwer, 1987.

[55] **Kruse, R., Schwecke, E., and Heinsohn, J.**, *Uncertainty and Vagueness in Knowledge Based Systems*, Springer-Verlag, Heidelberg, 1991.

[56] **Loo, S. G.**, Measures of fuzziness, *Cybernetica* 20, 201–210, 1977.

[57] **Lowen, R.**, Convex fuzzy sets, *Fuzzy Sets Systems* 3, 291–300, 1980.

[58] **Lubczonok, V.**, Fuzzy vector spaces, *Fuzzy Sets Systems* 38 (3), 329–343, 1990.

[59] **Mareš, M.**, How to handle fuzzy quantities? *Kybernetika*, 13 (1), 23–40, 1977.

[60] **Mareš, M. and Horák, J.**, Fuzzy quantities in networks, *Fuzzy Sets Systems* 10 (2), 123–134, 1983.

[61] **Mareš, M.**, Network analysis of fuzzy technologies, in *Fuzzy Methodologies for Industrial and System Engineering*, Evans, G. W., Karwowski, W., and Wilhelm, M. R., Eds., Elsevier, Amsterdam, 115–125, 1989.

[62] **Mareš, M.**, Some remarks to fuzzy CPM, *Ekonomicko-matematický obzor* 27 (4), 367–370, 1991.

[63] **Mareš, M.**, Addition of rational fuzzy quantities: Convolutive approach, *Kybernetika* 25 (1), 1–12, 1989.

[64] **Mareš, M.**, Addition of rational fuzzy quantities: Disjunction–conjunction approach, *Kybernetika* 25 (2), 104–116, 1989.

[65] **Mareš, M.**, Algebra of fuzzy quantities, *Int. J. General Systems* 20 (1), 59–65, 1991.

[66] **Mareš, M.**, Additive decomposition of fuzzy quantities with finite support, *Fuzzy Sets Systems* 47 (3), 341–346, 1992.

[67] **Mareš, M.**, Multiplication of fuzzy quantities, *Kybernetika* 28 (5), 337–356, 1992.

[68] **Mareš, M.**, Algebraic equivalences over fuzzy quantities, *Kybernetika* 29 (2), 121–132, 1992.

[69] **Mareš, M.**, Fuzzy data processing, *Fuzzy Sets Systems*, submitted.

[70] **Mareš, M.**, On fuzzification of multicriteria decision-making, *Fuzzy Systems A. I.*, 1 (2), 15–24, 1992.

[71] **Mareš, M.**, How to satisfy fuzzy manager, *Ekonomicko-matematický obzor* 24 (4), 396–404, 1988.

[72] **Mareš, M.**, A few remarks on vector optimization from coalition game theoretical point of view, *Kybernetika* 19 (4), 277–298, 1983.

[73] **Mareš, M.**, General coalition games, *Kybernetika* 14 (4), 245–260, 1978.

[74] **Mareš, M.**, Additivity of general coalition games, *Kybernetika* 14 (5), 350–368, 1978.

[75] **Mareš, M.**, Space of symmetric normal fuzzy quantities, *Int. J. General Systems*, submitted.

[76] **Mareš, M.**, Linear dependence of fuzzy vectors, in *Transactions of the International Symposium Fuzzy Approach to Reasoning and Decision Making*, Novák, V., Ramík, J., Mareš, M., Černý, M., and Nekola, J. (Eds.), Bechyně, June 25–29, 1990 (selected papers), Academia, Prague, 107–114, 1992.

[77] **Mareš, M.**, A few remarks on dependence, *BUSEFALL* 44, 6–8, 1990.

[78] **Mareš, M.**, Linear combination of fuzzy criteria, in *Proceedings of the International Workshop: Multicriteria Decision Making, Methods — Algorithms — Applications, Liblice, March 18 –22, 1991*, Černý, M., Glückaufová, G., and Loula, D., (Eds.), Inst. of Economics, Czechoslovak Academy of Sciences, Prague, 93–97, 1992.

[79] **Mareš, M.**, Remarks on fuzzy quantities with finite support, *Kybernetika*, 29 (2), 133–143, 1993.

[80] **Mareš, M.**, Equivalentions over fuzzy Quantities, in *Transactions of Second International Conference on Fuzzy Sets Theory and Its Applications*, Liptovský Mikuláš, January 31 –February 4, 1994, submitted.

[81] **Miyamoto, S.**, *Fuzzy Sets in Information Retrieval and Cluster Analysis*, Kluwer, Dordrecht, 1990.

[82] **Moore, R.**, *Interval Analysis*, Prentice Hall, Englewood Cliffs, 1966.

[83] **Nakamura, K.**, Preference relations on a set of fuzzy utilities as a basis for decision making, *Fuzzy Sets Systems* 20 (2), 147–162, 1986.

[84] **Nakamura, K.**, Canonical fuzzy numbers of dimension two and fuzzy utility difference for understanding preferential judgments, *Information Sciences* 50 (1), 1–22, 1990.

[85] **Negoita, C. V.**, *Fuzzy Systems*, Abacus Press, Kent, 1981.

[86] **Novák, V.**, *Fuzzy Sets and Their Applications*, A. Hilger, Bristol, 1989.

[87] **Novák, V. and Mareš, M. (Eds.)**, International Symposium on Fuzzy Approach to Reasoning and Decision Making, Bechyně, 1990 — abbreviated contributions, *Kybernetika*, supplement, 1–6, 1992.

[88] **Novák, V., Ramík, J., Mareš, M., Černý, M., and Nekola, J. (Eds.)**, Transactions of the International Symposium Fuzzy Approach to Reasoning and Decision Making, Bechyně, June 25–29, 1990 (selected papers), Academia, Prague, 1992.

[89] **Pawlak, Z.**, Rough sets, *Int. J. Comp. Inform. Sci.* 11 (3), 341–356, 1982.

[90] **Pawlak, Z.**, Rough Sets, *Theoretical Aspects of Reasoning about Data*, Kluwer, Dordrecht, 1991.

[91] **Pultr, A.**, Fuzziness and fuzzy equality, *Comment. Math. Univ. Carolina* 23, 249–267, 1982.

[92] **Ramík, J. and Římánek, J.**, Inequality relation between fuzzy numbers and its use in fuzzy optimization, *Fuzzy Sets Systems* 16, 123–138, 1985.

[93] **Ramík, J. and Římánek, J.**, Constrained optimization with fuzzy parameters, *Ekonomicko-matematický obzor* 21, 58–65, 1985.

[94] **Resconi, G., Klir, G. J., St. Clair, U., and Harmanec, D.**, On the integration of uncertainty theories, *Int. J. Uncertainty Fuzziness Knowledge-Based Systems*, 1, 1, 1993.

[95] **Rodabaugh, S.**, Fuzzy addition L-fuzzy real line, *Fuzzy Sets Systems* 8, 39–52, 1982.

[96] **Rosenfeld, A.**, Fuzzy groups, *J. Math. Anal. Appl.* 35, 512–517, 1971.

[97] **Roubens, M. and Vincke, Ph.**, *Preference Modelling*, Springer-Verlag, Berlin, 1985.

[98] **Ruspini, E. H.**, The semantics of vague knowledge, *Revue Internationale de Systemique*, 3 (4), 387–420, 1989.

[99] **Sanchez, E.**, Solution of fuzzy equations with extended operations, *Fuzzy Sets Systems* 12, 237–250, 1984.

[100] **Sanchez, E. and Gupta, M. M. (Eds.)**, *Proceedings of IFAC Symposium on Fuzzy Information, Knowledge Representation and Decision Processes*, Pergamon Press, Oxford, 1983.

[101] **Chen, S. J., Hwang, C. L., and Hwang, F. P.**, *Fuzzy Multiple Attribute Decision Making. Methods and Applications. Lecture Notes in Economics and Mathematical Systems 375*, Springer-Verlag, Berlin, 1992.

[102] **Toth, H.**, From fuzzy-set theory to fuzzy set-theory: Some critical remarks on existing concepts, *Fuzzy Sets Systems* 23, 219–237, 1987.

[103] **Tran Quoc Chien**, Medium distances of probability fuzzy-points and an application to linear programming, *Kybernetika* 25 (6), 494–504, 1989.

[104] **Vopěnka, P.**, *Mathematics Alternative Set Theory*, Teubner, Leipzig, 1979.

[105] **Wang, J. and Klir, G. J.**, *Fuzzy Measure Theory*, Plenum Press, New York, 1992.

[106] **Wang, P. P. and Chang, S. K. (Eds.)**, *Fuzzy Set Theory and Applications to Policy Analysis and Information System*, Plenum Press, New York, 1980.

[107] **Watson, S. R., Weiss, J. J., and Donnell, M. L.**, Fuzzy decision analysis, *IEEE Trans. Syst. Man. Cybernet.* 9, 1–9, 1979.

[108] **Yager R. R. (Ed.)**, *Fuzzy Sets and Applications: Selected Papers by L. A. Zadeh*, J. Wiley, New York, 1987.

[109] **Yager, R. R. and Zadeh, L. A. (Eds.)**, *An Introduction to Fuzzy Logic Applications in Intelligent Systems*, Kluwer, Boston, 1992.

[110] **Zadeh, L. A.**, Fuzzy sets, *Information Control*, 8 (3), 338–353, 1965.

[111] **Zadeh, L. A.**, Yes, no and relatively, *Chemtech*, June 1987, 340 – 344, and July 1987, 406–410.

[112] **Zadeh, L. A.**, Similarity relations and fuzzy-orderings, *Inf. Sci.* 3, 177–200, 1971.

[113] **Zadeh, L. A., Fu, K. S., Tanaka, K., and Shimura, M. (Eds.)**, *Fuzzy Sets and Their Applications to Cognitive and Decision Processes*, Academic Press, New York, 1975.

[114] **Zadeh, L. A. and Kacprzyk, J. (Eds.)**, *Fuzzy Logic for the Management of Uncertainty*, J. Wiley, New York, 1992.

[115] **Zimmermann, H. J.**, *Fuzzy Set Theory and its Applications*, Kluwer Nijhof, Boston, 1985.

[116] **Zimmermann, H. J.**, *Fuzzy Sets, Decision Making and Expert Systems*, Kluwer, Dordrecht, 1991.

Index

Q

R

S

T

U

V

W

Y

Z